Smart Grids for Renewable Energy Systems, Electric Vehicles and Energy Storage Systems

This comprehensive reference text discusses simulation with case studies and real-world applications related to energy system models, the large-scale integration of renewable energy systems, electric vehicles, and energy storage systems.

The text covers analysis and modeling of the large-scale integration of renewable energy systems, electric vehicles, and energy storage systems. It further discusses economic aspects useful for policy makers and industrial professionals. It covers important topics, including smart grids architectures, wide-area situational awareness (WASA), energy management systems (EMS), demand response (DR), smart grid standardization exertions, virtual power plants, battery degradation modeling, optimization approaches in modeling, and smart metering infrastructure.

The book:

- Discusses the analysis and modeling of the large-scale integration of renewable energy systems, electric vehicles, and energy storage systems.
- Covers issues and challenges encountered in the large-scale integration of electric vehicles, energy storage systems and renewable energy systems into future smart grid design.
- Provides simulation with case studies and real-world applications related to energy system models, electric vehicles, and energy storage systems.
- Discusses the integration of large renewable energy systems, with the presence of a large number of electric vehicles and storage devices/systems.

Discussing concepts of smart grids, together with the deployment of electric vehicles, energy storage systems and renewable energy systems, this text will be useful as a reference text for graduate students and academic researchers in the fields of electrical engineering, electronics and communication engineering, renewable energy, and clean technologies. It further discusses topics, including electric grid infrastructure, architecture, interfacing, standardization, protocols, security, reliability, communication, and optimal control.

Smart Grids for Renewable Energy Systems, Electric Vehicles and Energy Storage Systems

Edited by
Rajkumar Viral
Anuradha Tomar
Divya Asija
U. Mohan Rao
Adil Sarwar

CRC Press
Taylor & Francis Group
Boca Raton London New York

CRC Press is an imprint of the
Taylor & Francis Group, an **informa** business

First edition published 2023
by CRC Press
6000 Broken Sound Parkway NW, Suite 300, Boca Raton, FL 33487-2742

and by CRC Press
4 Park Square, Milton Park, Abingdon, Oxon, OX14 4RN

CRC Press is an imprint of Taylor & Francis Group, LLC

ISBN: 978-1-032-30095-5 (hbk)
ISBN: 978-1-032-31763-2 (pbk)
ISBN: 978-1-003-31119-5 (ebk)

DOI: 10.1201/9781003311195

Typeset in Sabon
by SPi Technologies India Pvt Ltd (Straive)

Contents

Preface

Smart grids are among the most momentous evolutionary changes in energy management systems because they support integrated systems, comprising renewable energy systems, the usage of large-scale renewable energy sources (RESs) and foremost developments in demand-side-management. Exploration into smart grid evolution has been realized globally for more than two decades, and thus there are now positive cases and extensive experience in this field. Similarly, electric vehicles (EVs) and modern energy storage systems (ESSs) epitomize one of the utmost encouraging technologies to green transportation, along with tremendous storage capability. The key concern is that the high level of infiltration of EVs leads to substantial electricity demand on the existing power grid. One potent way to lessen the impacts is to introduce locally available power generation such as RESs into the charging infrastructure. Because of the sporadic and in-dispatchable condition of RESs, it turn out to be very perplexing to manage the charging of EVs with other grid demands and renewable power generation. When the large-scale integration of EVs, RES and ESSs with the future smart grid there are numerous open issues and challenges counting the development and implementation, system security, system architectures, communication systems framework, standardization and protocols/guidelines, smart device interface, advanced forecasting, smart device interface, data transmission and monitoring, smart metering infrastructure, large charging infrastructure, operational and economical constraints etc. Consequently, seeing the future need and the requirement for a reliable and secure modern power system, this book intends to offer a deep insight into the aforementioned issue and challenges and for the large-scale integration of these three systems into future smart grid design.

Beginning with introductory chapters, which provide an overview of renewable energy technologies, smart grid technologies, and energy storage systems, the book progresses to the complexity of renewable energy integration with smart grid and the complicated control mechanisms.

Moreover, the book broadly covers the major threats, current and future trends in development and implementation, system security, architectural framework, various standardization levels, integrated system modelling and optimization methods, operational and economical aspects, and so on. There is an extensive discussion of the interactions of these systems with the smart grid as the future energy system model and methods and research solutions are considered for their potential in providing solutions in terms of large-scale integration.

The readers certainly get benefited with the concept of EVs, RESs, ESSs and smart grid covering basic philosophy, operational and implementation issues, system architecture in development, level of implementation, infrastructural challenges, standardization, etc.

It is our intention that this book will attempt to remove some of the highly contentious challenges in terms of major integration that the future smart grid poses, as well as their potential remedies.

About the Editors

Rajkumar Viral is presently working as an Associate Professor in Department of Electrical & Electronics Engineering, Amity School of Engg. & Tech., Amity University, Noida, UP, India, from August 2017 to the present day. Additionally, he has more than 14 years of teaching experience at undergraduate and postgraduate levels in various engineering institutes/universities. He was awarded a Doctoral degree (2010) and a Postgraduate degree in Energy Systems (2016) from Indian Institute of Technology Roorkee. He graduated in Electrical Engineering (with honors) from MJP Rohilkhand University, Bareilly in 2003. He has also published more than 100 research publications in various international/national journals/conferences/book chapters of repute and peer-reviewed and awarded with several Best Papers. He has served as TPC Member of several international conferences held across the globe. He also is an active member of professional societies i.e. the Institute of Electrical and Electronics Engineers, the Power & Energy Society, the Indian Society for Technical Education, the Institution of Engineers, and the American Council for an Energy-Efficient Economy, and contributing editorial board member (International Journal of Research in Computer Science, International Journal of Electrical and Computer Engineering, International Journal of BioSciences and Technology etc.) and a reviewer of various international journals (Elsevier, Springer, Taylor & Francis, Willey etc.). Her research interests include distribution system planning and optimizations, renewable energy systems & applications, smart grids, and Big Data analytics and their application in power systems.

Anuradha Tomar is currently working as Assistant Professor in Instrumentation & Control Engineering Division of Netaji Subhas University, Delhi, India. Dr. Tomar completed her Postdoctoral research in Electrical Energy Systems Group, from Eindhoven University of Technology (TU/e), the Netherlands and has successfully completed the European Commission's Horizon 2020, UNITED GRID and UNICORN TKI Urban Research projects. She has received her BE Degree in Electronics Instrumentation & Control with Honours from University of Rajasthan,

India and completed her MTech. Degree with Honours in Power System from the National Institute of Technology Hamirpur. She has received her PhD in Electrical Engineering from the Indian Institute of Technology Delhi (IITD). Her areas of research interest include the operation & control of microgrids, photovoltaic systems, renewable energy-based rural electrification, congestion management in LV distribution systems, artificial intelligence & machine learning applications in power systems, energy conservation and automation. She has authored or co-authored 69 research/review papers in various reputed international and national journals, and conferences. She is an editor for books with international publications such as Springer and Elsevier. She has also filed seven Indian patents in her name. Dr. Tomar is Senior member of the Institute of Electrical and Electronics Engineers, a Life member of Indian Society for Technical Education, The Institution of Electronics and Telecommunication Engineers, The Institution of Engineers, and International Association of Engineers

Divya Asija is an Assistant Professor in the Electrical and Electronics Engineering Department of Amity School of Engineering and Technology at the Amity University Noida, India, where she has been a faculty member since 2012. She has more than 14 years teaching experience at various prestigious institutions/universities. She completed his PhD at Amity University, Noida and her postgraduate studies at YMCA, Faridabad. Her research interests include power systems, ranging from congestion management to renewable energy resources, distributed generation, smart grids and electric vehicles. She has filed four patents and applied various funded projects of different government and private agencies. She is also the author of over forty research papers and book chapters in several international journals of repute She is also an active reviewer of international journals (*Energy* (Elsevier), the *International Journal of System Assurance Engineering and Management* (Springer), the International Journal of Emerging Electric Power Systems (Degruyter), the *International Journal of Electrical and Computer Engineering* (Institute of Advanced Engineering and Science), the *Journal of Electrical Engineering & Technology* (Springer) and the *International Journal of Renewable Energy Research*). In the interim, she was also awarded for outstanding contribution in reviewing by Energy, Elsevier for the academic years 2017–18 and 2018–19. As of 2020, Google Scholar reports over 121 citations to her work with h-index 6 and i10-index 4. She was also a winner of the gold award for courseware creation using Moodle for outcome-based education and ethical commitment.

U. Mohan Rao is a Senior Member, IEEE, and holds a bachelor's degree in electrical and electronics engineering, and a master's and doctoral degrees obtained from the National Institute of Technology, Hamirpur, India. At present, he is a lecturer in the Department of Applied Sciences at the

Université du Québec à Chicoutimi (UQAC) in Canada. He is also a post-doctoral researcher at UQAC with the Research Chair on the Aging of Power Network Infrastructure (ViAHT). Dr. Mohan is a Senior Member of IEEE and Member of the IEEE DEIS. He is also the Secretary for the IEEE Technical Committee on Liquid Dielectrics. His main research interests include aging phenomena of high-voltage insulation, condition monitoring of electrical apparatus, alternative dielectric materials, transformer insulation in cold countries, and AIML applications.

Adil Sarwar has received BTech., MTech., and PhD degrees from Aligarh Muslim University, Aligarh, India, in 2006, 2008, and 2012, respectively. Between 2012 and 2015, he was associated with the Electrical Engineering Department, Galgotia College of Engineering and Technology, Greater Noida, India. Since 2015, he has been working with the Department of Electrical Engineering, Aligarh Muslim University. He has also been a Senior Member of IEEE since 2020. He works in the area of the power converter and renewable energy systems. He has published more than 50 research papers in IEEE Transactions and SCI journals and he has also presented several papers in international conferences. Adil also chaired sessions in several international conferences and he has completed three NPIU-sponsored research projects. He is a Life Member of the Systems Society of India and has contributed a chapter to the *Power Electronics Handbook* (4th edn, edited by M. H. Rashid).

List of Contributors

Bilal Alam
Aligarh Muslim University, Aligarh, India

Wajid Ali
Aligarh Muslim University, Aligarh, India

Divya Asija
Amity University Uttar Pradesh, Noida, India

Mohd Bilal
Delhi Technological University, Delhi, India

Mohammed A. Bou-Rabee
College of Technical Studies, PAAET, Safat, Kuwait

V. Lakshmi Devi
Sri Venkateswara College of Engineering (SVCE), Tirupati, India

Hanumantha Reddy Gali
Sri Venkateswara Engineering College (SVEC), A.P., India

Anjali Jain
Amity University, Uttar Pradesh, Noida, India

Shivam Jain
Indian Institute of Technology Roorkee, Uttarakhand, India

Supriya Jaiswal
National Institute of Technology, Hamirpur, India

M. Jayachandran
Sri Manakula Vinayagar Engineering College, Puducherry, India

C. Kalaiarasy
Puducherry Technological University, Puducherry, India

C. Kalaivani
Puducherry Technological University, Puducherry, India

Azam Khan
Aligarh Muslim University, Aligarh, India

Chiranjib Koley
NIT, Durgapur, India

Kumar K.
Sri Venkateswara College of Engineering (SVCE), A.P. India

Reinhard Madlener
Norwegian University of Science
and Technology (NTNU),
Trondheim, Norway

Ashish Mani
Amity University, Uttar Pradesh,
Noida, India

Safwan Mustafa
Aligarh Muslim University, Aligarh,
India

Anita Devi Ningthoujam
Manipur Technical University,
Manipur India

Amruta Pattnaik
Dr Akhilesh Das Gupta Institute
of Technology & Management,
India

Avagaddi Prasad
Sasi Institute of Technology &
Engineering, A.P., India

Khaliqur Rahman
Aligarh Muslim University, Aligarh,
India

M. Rizwan
Delhi Technological University,
Delhi, India

Felipe Sabadini
Aachen University, Aachen,
Germany

Sohit Sharma
Tata Consultancy Services, Pune,
India

Anwar S. Siddiqui
Jamia Milia Islamia, New Delhi,
India

A. Pullabhatla Srikanth
NIT, Durgapur, India

Mohd Tariq
Aligarh Muslim University, Aligarh,
India

Ramji Tiwari
Sri Krishna College of Engineering
and Technology, T.N India

Anuradha Tomar
Netaji Subhas University of
Technology, India

Rajkumar Viral
Amity University Uttar Pradesh,
Noida, India

Chapter 1

Introduction to E-vehicle technology

Amruta Pattnaik

Dr Akhilesh Das Gupta Institute of Technology & Management,
New Delhi, India

Anuradha Tomar

Netaji Subhas University of Technology, India

CONTENTS

1.1 INTRODUCTION

The electric vehicle (EV) is a vehicle that is operated either by electric motors and batteries or by solar photovoltaic cells. E-vehicles are free from combustion engines. They currently involve a combination of a converter along with the controller, batteries, and cooling process as well as the transmission system. Such components do not create any type of emissions, making them environment-friendly. E-vehicles are becoming increasingly popular as it eliminates pollution, saves natural resources, and has other advantage [1, 2].

Historically, e-vehicles have been powered by a battery. Afterwards, other types of e-vehicles were introduced based upon an amalgam of liquid, gaseous, or fuel-based power-driven energy. One of the earliest innovators was Anyos Jedlik, who designed an electric motor for vehicles [1]. Robert Anderson, a Scot, invented the first simple electric bearing operated by non-rechargeable batteries [1]. In 1835, Sibrandus Stratingh from the Netherlands had invented a non-rechargeable electrical vehicle. Later, William Morrison developed the first electric vehicle, which covered a distance of 14 miles/hour and was also able to carry 6 people [2]. In 1859, a primary cell batteries (i.e. a non-rechargeable battery) was used in e-vehicles. Non-rechargeable batteries have some disadvantages, however, such as the generation of electric waste, high cost due to their bulky size, and the need for a large number of batteries to power just a single e-vehicle. The Studebaker Automobile Company provided the first electric vehicle, which was operated by gas.

DOI: 10.1201/9781003311195-1

Similar developments during this time were to be observed in electric trains, which are both popular and able to operate at relatively high speeds; they are also economical [2]. It is generally agreed that the first practical electric car was created by Thomas Parker in 1884. In 1899, another electric car, designed by the renowned designer, Ferdinand Porsche, was manufactured in Germany. Moving to the present day, China is now providing a good platform for EVs [3, 4]. By 1920, the EV market was already achieving huge success as 28% of EVs were manufactured in the U.S. Then, however, Henry Ford established the Model T with a new mass production process [5–10].

In the 21st century, different types of electric vehicles have been introduced to overcome the issue of pollution as well as to replace petrol/diesel-based vehicles due to the shortage of fossil fuels. Among the major types of e-vehicles are: (i) battery e-vehicles; (ii) hybrid e-vehicles; (iii) plug-in hybrid e-vehicles; (iv) range-extended e-vehicles; (v) fuel cell e-vehicles; and (vi) solar-based e-vehicles. The major advantages of and e-vehicles are as follows [11, 12]:

   (i) It reduces the emission of greenhouse gases
  (ii) It does not require petroleum fuels or gas to start the vehicle
 (iii) It reduces pollution
 (iv) It shows good performance on urban roads.
  (v) It reduces maintenance costs
 (vi) There is no noise pollution
(vii) It is lighter, smoother, and faster than vehicles powered by the internal combustion engine
(viii) It is quite convenient to drive.

The major disadvantage of e-vehicles at present is the requirement for an infrastructure of charging stations for e-vehicles and the charging time for batteries are quite long. A long charging time causes large amounts of electricity. Electricity Storage are another drawback of e-vehicles. Lack of commercialization and limited speed range are the other disadvantages of e-vehicles [13].

This essay contains four further sections. Section 1.2 will discuss the different types of e-vehicles. Section 1.3 will report about the recent technologies in electric vehicles and section 1.4 will give an idea about the standards which are used for electric vehicles. Finally, section 1.5 offers some conclusions to the discussion.

1.2 DIFFERENT TYPES OF E-VEHICLES

As has already been mentioned in the above section, there are several different types of electric vehicles on the market. Among the categories of major e-vehicles are the following [14]:

1. Battery e-vehicle [BEV]: these types of vehicle are operated by batteries. In addition, they are free from an internal combustion engine and oil tank. Such vehicles require access to an external electrical charging station. In some cases, the process of regenerative braking can recharge the batteries [15].

2. Plug-in hybrid e-vehicle [PHEVs]: Vehicles which can use either gasoline or electricity are referred to as plug-in hybrids. They do not require an external electrical charging station since this is an integral part of the vehicle. They need a gas engine for both the charging and range extension [16].

3. Hybrid e-vehicle: This is operated by both fuel (gasoline) and electricity. It can recharge the battery through the regenerative technique as well as the fuel. It needs an electric motor for driving purposes, which is powered by a fuel engine. It does not require an external electrical charging station too as it is an integral part of the vehicle [17].

4. Fuel cell e-vehicle: The fuel cell is one of the non-conventional energy sources which provides electricity from hydrogen and oxygen. Fuel cell-based e-vehicles are the most commonly used. The advantage of this vehicle is based upon the plentiful supply of H_2O. Such vehicles are currently still in the development phase. The fuel cell e-vehicle the Toyota Mirai was introduced in 2015 [18].

5. Solar cell-based e-vehicle: such vehicles use solar energy to recharge their batteries. They are able to use their power to increase their speed. In addition, they are eco-friendly, give off zero emissions, and require less maintenance, in addition to having other advantages. They are quite costly, however, and also depend upon the position of the Sun as well as the efficiency of their storage batteries [19].

6. Range- extended e-vehicle: This is operated by high-powered batteries. It requires a combustion engine to recharge the battery [20].

1.3 TECHNOLOGIES SO FAR

Electric vehicles are eco-friendly vehicles. Today governments are trying to encourage manufacturing companies into developing and producing such vehicles. E-vehicles has advantages such as their eco-friendliness, their low (or zero) fuel costs, etc. The major disadvantages of the e-vehicle at present are the charging stations and storage batteries, etc. [21]. Different types of technologies have been introduced, but three different main technologies are discussed here. These are as follows [4, 22]:

1. Batteries for e-vehicles
   Traction batteries are used to provide energy to propulsion systems in the electric vehicle. The role of batteries is not only to operate the electric vehicle but also to ensure that power should be provided

continuously. Accordingly, the battery has the capacity for large stored energy. Following the introduction of lead-acid batteries, different other types of batteries have been introduced onto the electric vehicle market [23]. These are as follows:

a. Pb-acid batteries
   - First rechargeable battery i.e., Pb-acid battery, was invented by French expert Dr. Planté in 1859 [24].
   - They are mostly used in moving and immobile applications
   - The Pb-acid battery contains Pb, PbO_2, and H_2So_4
   - The size and geometry of the electrodes in Pb-acid batteries are responsible for the size of power and energy
   - The advantages of Pb-acid batteries are that they are lower prices, high energy, and easily recyclable.
   - The Pb-acid batteries are not eco-friendly due to the presence of lead, so it is quite unsafe for disposal or recycling.
   - Pb-acid batteries are less expensive than Li-ion batteries
   - Due to a lack of sufficient specific energy and density, Pb-acid batteries are not useful in many electric vehicles.

b. Ni-metal hydride batteries
   - First introduced in Japan in 1990 [25].
   - The batteries include positive Ni-hydroxide, negative metal hydride, and nylon separator sheets.
   - Environmentally friendly, low price, and energy density is higher.
   - The Ni-MH battery has only half the energy of a Li-ion battery.
   - Safety, performance, and durability is affected due to fast charge causes heat and high-load discharge.

c. Lithium-ion batteries
   - It is used in wireless phones and computers due to its high energy/unit mass as compared to energy storage systems [26]
   - It acts very well in high temperatures and is highly efficient,
   - It has a high P/W [power to weight] ratio and low auto-release charge.
   - Li-ion batteries can be recycled,
   - In industry, it is difficult to recover the cost of material from the recycling of the battery.

d. Other batteries
   Na-NiCl$_2$ batteries are also called ZEBRA batteries. The "Zero Emission Battery Research Activity" (ZEBRA) project established the concept of the ZEBRA battery. The merits of these batteries are maximum energy density, low deterioration, high protection, less sensitivity to excess charge and release of charge, good lifetime, and the low price for EV batteries. The disadvantages of such batteries are low specific power, self-discharge issue and excess heat requirement. They are widely used in submarines, defense applications, wire facilities, and energy storage [27–29].

Metal-air batteries contains two electrodes. The positive electrode is made up of metal and the air is acted as the negative electrode. Different types of metals, such as Li, Ca, Mg, Fe, Al, or Zn, are used as the anode. Lithium-air (Li-air) batteries are effective [30]. Sodium-beta batteries use solid-state electrolytes and are categorized into sodium-sulfur (Na–S) and sodium metal halide types. Ultracapacitors (UCs) can achieve high power in the range of 10^3–2×10^3 W/kg with an efficiency of 95%. They can use the energy from the braking method and vary the speed of EVs due to their ability both to rapidly charge and discharge [31].

2. Charging Technology

Charging technology is one of the most important aspects of electrical vehicles. There is an important interrelationship between the charging and the battery technology. Charging is becoming easier and more rapid due to the spread of charging stations and charging technology. There are three different modes of charging, as shown in Figure 1.1 [32]:

(1) Inductive
(2) Conductive and
(3) Swapping.
(1) Inductive charging:
   (a) This is also known as wireless charging [33–35].
   (b) In this wireless charging concept, the energy is transferred through an air gap from one coil to another.
   (c) It is easy to install at airports and gas stations.
   (d) Charging can also be achieved during driving.

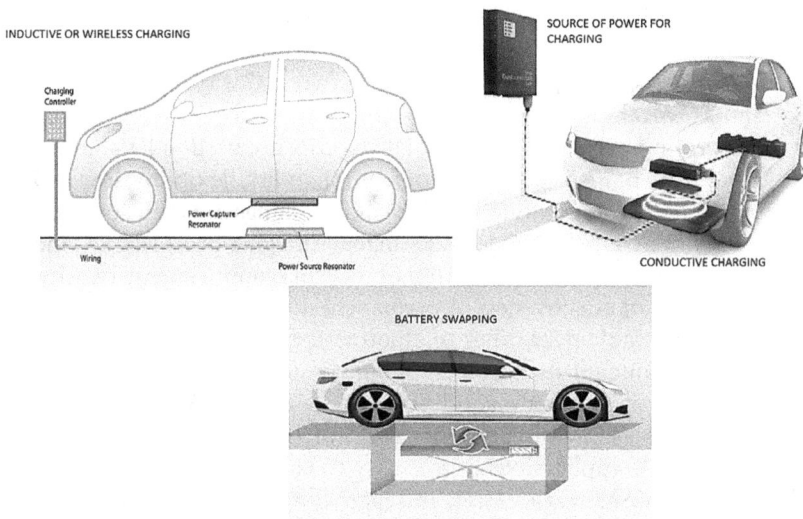

Figure 1.1 Different types of charging technology.

(e) There are two types, i.e. inductive power transfer and capacitive power transfer.

(f) Recently, a charging company called Char.gy (which was launched in the UK in October 2021) has begun a trial on wireless charging using 10 wireless charging pads connected into car parking spaces across the county of Buckinghamshire.

(2) Conductive:

(a) It requires a physical connection of charger and motor [36, 37].

(b) It requires an AC, DC, AC/DC, or Dc/DC converter as per the need for charging.

(c) Types of the conductive charger: (i) on board: it is applicable for slow charging and carried out inside the vehicle. (ii) off-board charger: It is used for DC fast charging at level 3. Those are placed at the urban area and a highway refuelling station

(d) The required power for charging can be drawn from the solar cell, grid, or battery.

(3) Battery Swapping

(a) It is an easy method for charging [38].

(b) It needs to swap the exhausted battery with the new one at the battery swapping station.

(c) It provides a payable service to provide a chargeable battery

(d) The battery exchange place includes a transformer, rectifiers/inverters, cell chargers, mechanical arms, indicting stands, care systems, regulator systems, and other apparatus included in the exchange and charging of the cells [39].

3. Motor technology

Electric vehicles require electric motors which are placed at the propulsion system. The motor converts stored electrical energy into kinetic energy for the vehicle. There are two major types of motors, DC and AC machines [40]. Both asynchronous motor switched reluctance motors and permanent magnet DC motors are used in EVs. Induction motors are quite popular due to their reliability, flexibility, low maintenance, and relative value. The key shortcomings are their low levels of efficiency at small loads [41]. Permanent magnet brushless DC motors are implemented in EVs because of their maximum energy density and high levels of effectiveness. A permanent magnet synchronous motor is quite preferable over other AC motors for their propulsion system as it has advantages such as their simple design, reliability, high efficiency as well as relatively good power density [42]. A surface-mounted Permanent Magnet Synchronous Motor (PMSM) is quite frequently used for EV applications as it is easy to control when compared with interior PMSM. Cost, noise, and high vibration are among the disadvantages of PMSM. Another motor, labelled the switched reluctance motor, is also used in EV application due to its efficiency, low price, ease of control, and simple structure. Most of the traction type motors are used in EV applications [43].

1.4 STANDARDS FOR E-VEHICLES

According to the 'Vehicle Codes and Standards: Overview and Gap Analysis' by C. Blake, W. Buttner, and C. Rivkin (Technical Report NREL/TP-560-47336 February 2010), codes and standards for e-vehicles are decided at both the national and world market levels. Such types of standards are mentioned by a committee of both interested parties and shareholders. Many codes and standards technical committees are led by the National Renewable Energy Laboratory (NREL) and NREL also monitors the codes and standards technical committees' actions. The National Fire Protection Association (NFPA), the International Organization for Standardization (ISO), the Society of Automotive Engineers (SAE), and the ISO Draft International Standard (DIS) are included with the NREL to develop the codes and standard process [44].

Standards are designed so that the customer is able to receive consistent, dependable, and safe mode operation from machinery. Standards for e-vehicles confirm that there will be standardization of any Electric Vehicle Supply Equipment (EVSE) across all e-vehicles [45–47].

At the international level, there are different organizations to develop the standards for e-vehicles. The Society of Automotive Engineers (SAE) Standards have been quite popular for e-vehicle standards. SAE standards are used for: (i) light vehicle exterior sound level and standards; (ii) light-duty vehicle performance and economy measures; (iii) truck and bus and hybrid and e-vehicles; (iv) truck and bus alternative fuels; (v) fuel system standards; (vi) hybrid standards; (vii) fuel cell standards; (viii) storage battery standards; and (ix) engine power standards. The Central Electricity Authority (CEA) and the Bureau of Indian Standards (BIS) are the national standards of India, being used for electric vehicle (EV) charging standards. The Automotive Research Association of India (ARAI) has also been involved in developing standards for vehicles and their components [47].

Current codes and standards are quite helpful for policy makers in terms of either developing or hampering the electric vehicle system's equipment distribution and approval. Before providing any type of codes and standards, the committee is set a regulatory framework, as shown in Figure 1.2.

Before any replacement or modification in codes and standards can be introduced, the committee must review the existing regulatory framework. After completing the study of an existing regulatory framework, codes will

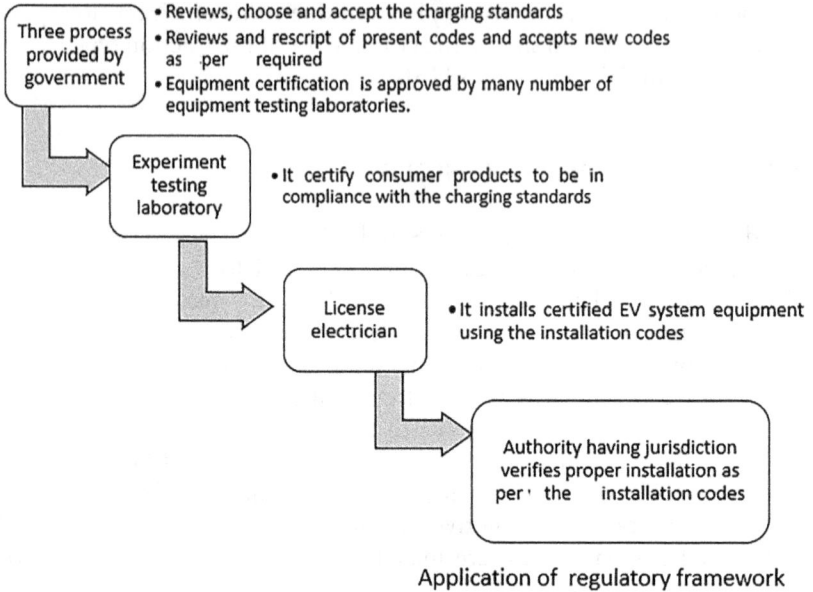

Application of regulatory framework

Figure 1.2 Regulatory framework for codes and standards [46].

be modified and charging standards will be adopted under the consideration of common existing standards at the international level as well as types of charging standards. Adopting the charging standards eases the movement of e-vehicles between countries [47]. Some of the codes and standards are summarized below:

- Installation codes:
  - National Fire Protection Association's National Electric Code [47]
    - Accepted by North America and certain South American countries
    - connection of circuits and equipment
    - used for homes and commercial buildings
    - used especially in electric vehicle system equipment
  - International Code Council: International Building Code and International Residential Code [47]
    - International Building Code used in the United States, the Middle East, and the Caribbean (ICC 2015).
    - Connection of electrical circuits and equipment
    - Applies to both commercial and residential settings.
    - This code includes requirements concerning circuitry, power systems, and power supply that apply to EVSE installation
  - Standard codes [47]:

Level 1   and Level 2 couplers (AC charging) used for IEC 62196

Type 1   Type 2   GB/T (AC)

Mennekes connector   SAEJ1772 Connector   it is a connector in china

Figure 1.3 Level 1 and level 2 couplers.

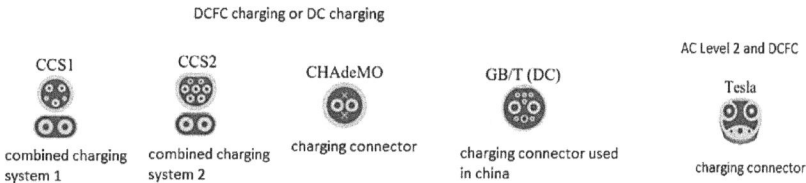

DCFC charging or DC charging

AC Level 2 and DCFC

CCS1   CCS2   CHAdeMO   GB/T (DC)   Tesla

combined charging   combined charging   charging connector   charging connector used   charging connector
system 1   system 2   in china

Figure 1.4 DCFC charging.

- IEC 62196 Standard
  - Used for charging connectors
  - Level 1, Level 2, and DCFC connectors are registered in the IEC 62196, as shown in Figures 1.3 and 1.4.
  - Either Type 1 or Type 2 for the Level 1 and Level 2 charging standards are the popular choices for countries.
  - Type 1 is widely used in North America and Japan
  - Type 2 can be applied for Level 1 and Level 2 charging. This connector is most common in the European Union.
  - DCFC connectors are Combined Charging System 1 (CCS1), Combined Charging System 2(CCS2), CHAdeMO (only be used for DCFC), and GB/T (connector used for DC charging).
  - AC level 2 and DCFC are used in Tesla, a charging connector, used except in China and Europe.
- IEC 61851 Standard [47]
  - It is used for electric vehicle system equipment. It has four modes, as shown in Figure 1.5.
  - Based on the working voltage, the active power, with or without fault, and safety features.
  - In-cable control and protection device (IC-CPD) are placed in a cable to protect from fault
  - The IEC 61851 standard distinguishes the position of fault, as shown in Figure 1.5.

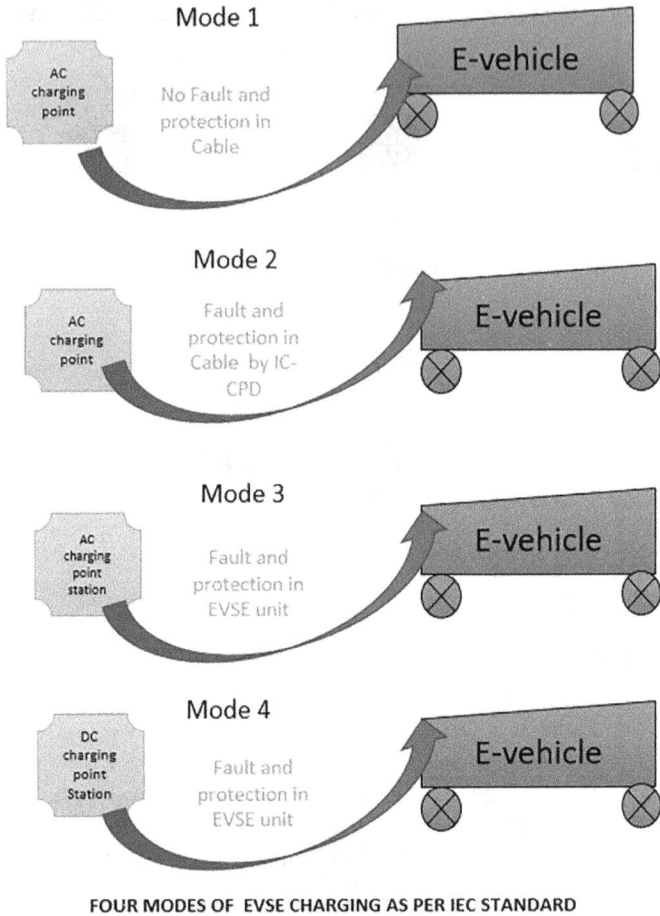

FOUR MODES OF EVSE CHARGING AS PER IEC STANDARD

Figure 1.5 Four modes of charging.

- Mode 1: Charging an EV from a standard electrical outlet without IC-CPD in the cable. This mode is not adopted in some countries as it has no protection structures.
- Mode 2: Charging an EV with an IC-CPD.
- Mode 3: Charging an EV from a permanently installed AC electric vehicle supply equipment in either a residential or public setting by an IC-CPD.
- Mode 4: Charging an EV from a permanently installed DC electric vehicle supply equipment by an IC-CPD cable in public places.

Here, it has discussed the above two standards for e-vehicles. Some of the other specific charging standards are listed in Figure 1.6.

IEC 62196 Standard	Electrical specifications and design of several configurations of charging connectors developed by different manufacturers
IEC 61851 Standard	four modes of EVSE and categorizes each by the operating voltage, the power (i.e., kW) delivered to the EV, and the presence or absence of fault and protection features
International Organization for Standardization 15118	An open payment and communication standard that allows for secure and automatic payment from a driver's account once a driver plugs the EVSE into the EV. This standard is common in Europe
J-1773	SAE Electric Vehicle Inductively-coupler charging
J-2954	Wireless power transfer for light-duty plug-in Electric vehicles and alignment methodology
62196-x	Plug, Socket-Outlets, Vehicle Couplers and Vehicle inlets–Conductive charging of electric vehicles
61851	61851-X Electric vehicles conductive charging system
62840	62840-X Electric vehicle battery swap system
62983	62983 Electric charge station
61980-1:2015	Electric vehicle wireless power transfer systems
GB/T 18487.X-2015	Electric J-1772c vehicles conductive charging system
GB/T 20234.X-2015	Electric vehicles conduction charging connecting device
GB/T 27930-2015	Communication protocol between the off-board charger and battery management system
QC/T 895-2011	Electric vehicles onboard charger
QC/T 841-2010	Electric vehicles conductive charging interface
C601: 2000	Plugs and receptacles for EV charging
G105: 1993	Connectors applicable to quick charging system at Eco-Station for EVs
G106: 2000	EV Inductive charging system: General requirements
G107: 2000	Inductive charging system: Manual connection
G108: 2001	EV Inductive charging system: Software interface
G109: 2001	EV Inductive charging system: General requirements
	X: defines multiple standards

Figure 1.6 Arraigning ethics based on IEC, SAE, ICE, GB, and JEVS [46, 47].

1.5 CONCLUSION

This work has provided a very brief review of electrical vehicles. It focuses on the different types of technology used in electrical vehicles as well as the international standards used in electrical vehicles. Among different types of EVs, both solar and fuel cell-based EVs are emerging trends in electric vehicles. Renewable energy-based EVs are eco-friendly, and have zero missions as well as zero fuel costs.

The battery is an integral part of EVs, with lead-acid batteries currently in widespread use. Presently, Li-ion batteries and supercapacitors have replaced applications using conventional batteries. Among different types of charging methods, conductive charging is popular due to a lack of awareness and commercialization. Induction motors have proved useful in EVs. The power electronics-based induction motor is quite good for EVs as it has good capabilities into the extension of range.

For EV technology, some of the standards and codes are fixed by the NREL to provide good service for customers. IEC and SAE usually provide the codes and standards for EVs. At present, EVs have technical advantages as well as disadvantages, but new technologies are being introduced continuously to promote the spread of EV use across society. In the future, EVs will have completely replaced all diesel or gasoline vehicles.

REFERENCES

1. Jones, W.D., 2003. Hybrids to the rescue [hybrid electric vehicles]. *IEEE Spectrum, 40*(1), pp. 70–71.
2. Jones, W.D., 2005. Take this car and plug it [plug-in hybrid vehicles]. *IEEE Spectrum, 42*(7), pp. 10–13.
3. Giripunje, L.M., Singh, V.K. and Kendre, G.S., 2021. A review of dual-motor system and methods of charging batteries of an e-vehicle. *EPRA International Journal of Research & Development, 6*(6), pp. 124–126.
4. Gupta, S., Electric Vehicle Charging the Future with Batteries (deep techexpress.com)[Published December 2, 2021].
5. Sun, X., Li, Z., Wang, X. and Li, C., 2020. Technology development of electric vehicles: A review. *Energies, 13*(1), p. 90.
6. Li, W., Long, R. and Chen, H., 2016. Consumers' evaluation of national new energy vehicle policy in China: An analysis based on a four paradigm model. *Energy Policy, 99*, pp. 33–41.
7. Hu, Z. and Yuan, J., 2018. China's NEV market development and its capability of enabling premium NEV: Referencing from the NEV market performance of BMW and Mercedes in China. *Transportation Research Part A: Policy and Practice, 118*, pp. 545–555.
8 Du, J. and Ouyang, D., 2017. Progress of Chinese electric vehicles industrialization in 2015: A review. *Applied Energy, 188*, pp. 529–546.
9. Guarnieri, M., 2012. September. Looking back to electric cars. In *2012 Third IEEE HISTory of ELectro-technology CONference (HISTELCON)* (pp. 1–6). IEEE.
10. Chan, C.C., 2012. The rise & fall of electric vehicles in 1828–1930: Lessons learned [scanning our past]. *Proceedings of the IEEE, 101*(1), pp. 206–212.
11. Shahan, Z., 2016. Million pure EVs worldwide: EV revolution begins. *Clean technica, pages the kidwind project: Using mini-supercapacitors.*
12. Hannan, M.A., Lipu, M.H., Hussain, A. and Mohamed, A., 2017. A review of lithium-ion battery state of charge estimation and management system in electric vehicle applications: Challenges and recommendations. *Renewable and Sustainable Energy Reviews, 78*, pp. 834–854.
13. Pelegov, D.V. and Pontes, J., 2018. Main drivers of battery industry changes: Electric vehicles—a market overview. *Batteries, 4*(4), p. 65.
14. Saxena, S., Le Floch, C., MacDonald, J. and Moura, S., 2015. Quantifying EV battery end-of-life through analysis of travel needs with vehicle powertrain models. *Journal of Power Sources, 282*, pp. 265–276.
15. Alshahrani, S., Khalid, M. and Almuhaini, M., 2019. Electric vehicles beyond energy storage and modern power networks: Challenges and applications. *IEEE Access, 7*, pp. 99031–99064.

16. Nykvist, B. and Nilsson, M., 2015. Rapidly falling costs of battery packs for electric vehicles. *Nature Climate Change*, 5(4), pp. 329–332.

17. Reddy, T.B., 2011. *Linden's handbook of batteries*.4th Edition. New York: McGraw-Hill Education.

18. Horkos, P.G., Yammine, E. and Karami, N., 2015, April. Review on different charging techniques of lead-acid batteries. In *2015 Third International Conference on Technological Advances in Electrical, Electronics and Computer Engineering (TAEECE)* (pp. 27–32). IEEE. Beirut, Lebanon. DOI: 10.1109/TAEECE.2015.7113595

19. May, G.J., Davidson, A. and Monahov, B., 2018. Lead batteries for utility energy storage: A review. *Journal of Energy Storage*, 15, pp. 145–157.

20. Tie, S.F. and Tan, C.W., 2013. A review of energy sources and energy management system in electric vehicles. *Renewable and Sustainable Energy Reviews*, 20, pp. 82–102.

21. Chau, K.T., Wong, Y.S. and Chan, C.C., 1999. An overview of energy sources for electric vehicles. *Energy Conversion and Management*, 40(10), pp. 1021–1039.

22. Li, G., Lu, X., Kim, J.Y., Meinhardt, K.D., Chang, H.J., Canfield, N.L. and Sprenkle, V.L., 2016. Advanced intermediate temperature sodium–nickel chloride batteries with ultra-high energy density. *Nature Communications*, 7(1), pp. 1–6.

23. Hannan, M.A., Hoque, M.M., Mohamed, A. and Ayob, A., 2017. Review of energy storage systems for electric vehicle applications: Issues and challenges. *Renewable and Sustainable Energy Reviews*, 69, pp. 771–789.

24. Khaligh, A. and Li, Z., 2010. Battery, ultracapacitor, fuel cell, and hybrid energy storage systems for electric, hybrid electric, fuel cell, and plug-in hybrid electric vehicles: State of the art. *IEEE Transactions on Vehicular Technology*, 59(6), pp. 2806–2814.

25. Chen, A. and Sen, P.K., 2016, October. Advancement in battery technology: A state-of-the-art review. In *2016 IEEE Industry Applications Society Annual Meeting* (pp. 1–10). IEEE.

26. Shen, Y., Noréus, D. and Starborg, S., 2018. Increasing NiMH battery cycle life with oxygen. *International Journal of Hydrogen Energy*, 43(40), pp. 18626–18631.

27. Manzetti, S. and Mariasiu, F., 2015. Electric vehicle battery technologies: From present state to future systems. *Renewable and Sustainable Energy Reviews*, 51, pp. 1004–1012.

28. Ding, Y., Cano, Z.P., Yu, A., Lu, J. and Chen, Z., 2019. Automotive Li-ion batteries: Current status and future perspectives. *Electrochemical Energy Reviews*, 2(1), pp. 1–28.

29. Bresser, D., Hosoi, K., Howell, D., Li, H., Zeisel, H., Amine, K. and Passerini, S., 2018. Perspectives of automotive battery R&D in China, Germany, Japan, and the USA. *Journal of Power Sources*, 382, pp. 176–178.

30. Lee, J.H., Yoon, C.S., Hwang, J.Y., Kim, S.J., Maglia, F., Lamp, P., Myung, S.T. and Sun, Y.K., 2016. High-energy-density lithium-ion battery using a carbon-nanotube–Si composite anode and a compositionally graded Li [Ni 0.85 Co 0.05 Mn 0.10] O 2 cathode. *Energy & Environmental Science*, 9(6), pp. 2152–2158.

31. Akhil, A.A., Huff, G., Currier, A.B., Kaun, B.C., Rastler, D.M., Chen, S.B., Cotter, A.L., Bradshaw, D.T. and Gauntlett, W.D., 2015. DOE/EPRI electricity storage handbook in collaboration with NRECA. *Sandia National Laboratories*.

32. Nazri, G.A. and Pistoia, G. eds., 2008. *Lithium batteries: Science and technology*. Springer Science & Business Media.

33. Yoo, H., Sul, S.K., Park, Y. and Jeong, J., 2008. System integration and power-flow management for a series hybrid electric vehicle using supercapacitors and batteries. *IEEE Transactions on Industry Applications*, *44*(1), pp. 108–114.

34. Haddoun, A., Benbouzid, M.E.H., Diallo, D., Abdessemed, R., Ghouili, J. and Srairi, K., 2007. A loss-minimization DTC scheme for EV induction motors. *IEEE Transactions on Vehicular Technology*, *56*(1), pp. 81–88.

35. Gan, J., Chau, K.T., Chan, C.C. and Jiang, J.Z., 2000. A new surface-inset, permanent-magnet, brushless DC motor drive for electric vehicles. *IEEE Transactions on Magnetics*, *36*(5), pp. 3810–3818.

36. Chau, K.T., Chan, C.C. and Liu, C., 2008. Overview of permanent-magnet brushless drives for electric and hybrid electric vehicles. *IEEE Transactions on Industrial Electronics*, *55*(6), pp. 2246–2257.

37. Rahman, K.M., Fahimi, B., Suresh, G., Rajarathnam, A.V. and Ehsani, M., 2000. Advantages of switched reluctance motor applications to EV and HEV: Design and control issues. *IEEE Transactions on Industry Applications*, *36*(1), pp. 111–121.

38. Jones, W.D., 2007. Putting electricity where the rubber meets the road [NEWS]. *IEEE Spectrum*, *44*(7), pp. 18–20.

39. Affanni, A., Bellini, A., Franceschini, G., Guglielmi, P. and Tassoni, C., 2005. Battery choice and management for new-generation electric vehicles, *IEEE Transactions on Industrial Electronics*, *52*(5), pp. 1343–1349.

40. Chan, C.C. Present status and future trends of electric vehicles. In *1993 2nd International Conference on Advances in Power System Control, Operation and Management, APSCOM-93*, pp. 456–469. IET, 1993.

41. Mahoor, M., Hosseini, Z.S. and Khodaei, A., 2019. Least-cost operation of a battery swapping station with random customer requests. *Energy*, *172*, 913–921.

42. Sultana, U., Khairuddin, A.B., Sultana, B., Rasheed, N., Qazi, S.H. and Malik, N.R. 2018. Placement and sizing of multiple distributed generation and battery swapping stations using grasshopper optimizer algorithm. *Energy*, *165*, 408–421.

43. Zheng, J., Mehndiratta, S., Guo, J.Y. and Liu, Z., 2012. Strategic policies and demonstration program of electric vehicle in China. *Transport Policy*, *19*, 17–25.

44. Scott, H., 2019. Understanding the impact of reoccurring and non-financial incentives on plug-in electric vehicle adoption—A review. *Transportation Research Part A: Policy and Practice*, *119*, 1–14.

45. Hao, H., Ou, X., Du, J., Wang, H. and Ouyang, M. 2014. China's electric vehicle subsidy scheme: Rationale and impacts. *Energy Policy*, *73*, 722–732.

46. Zhang, X., Liang, Y., Yu, E., Rao, R. and Xie, J., 2017. Review of electric vehicle policies in China: Content summary and effect analysis. *Renewable and Sustainable Energy Reviews*, *70*, 698–714.

47. Aznar, Alexandra, Belding, Scott, Bopp, Kaylyn, Coney, Kamyria, Johnson, Caley and Zinaman, Owen, 2021. *Building blocks of electric vehicle deployment a guide for developing countries*. No. NREL/TP-7A40-78776. National Renewable Energy Lab.(NREL), Golden, CO (United States), 2021.

Chapter 2

Electric vehicles and smart grid interactions

Infrastructure, current trends, impacts and challenges

Supriya Jaiswal
NIT Hamirpur, Hamirpur, India

Sohit Sharma
Tata Consultancy Services, Pune, India

CONTENTS

2.1 INTRODUCTION

Electric Vehicles (EVs) have gained increased admiration and acceptance in the upcoming green transportation era. The major reason behind the increased utilization of EVs is to reduce the harmful emission produced by the internal combustion engine (ICE)-based vehicles, i.e. NMOG by 98%, NO_x by 92% and CO by 99% [1]. In addition, the limited accessibility of fossil fuels have forced the deployment of EVs in the world. Due to the large

DOI: 10.1201/9781003311195-2

volume of ICE-based vehicles, the air quality of India has been ranked third with degraded air quality index 141 among 103 countries around the globe in 2020 [2]. As a result, India is committed to reduce greenhouse gases emission by 33% to 35% below the 2005 level by 2030, using zero emission vehicles [3]. In an attempt to control air pollution, in many countries policies to end the sale of new ICE vehicles have been introduced, e.g. Norway has planned to implement this plan by 2025, to be followed by Denmark, the UK and Ireland by 2030. EVs have now entered their third century as a commercially available transportation mode and they remain successful in its implementation, usage, market sales and lower environmental impacts when compared with other transportation-based technical solutions. In order to fully understand the principle behind the design of EVs, including their types, components, structure and relevant technological and environmental aspects, it is important to revisit the long history of EVs [4].

2.1.1 Historic timelines

19th century – the Beginning of EVs: The initial small-scale electric cars were first invented in the Netherlands, the USA and Hungary in the years 1828–1835, in a period when the primary mode of transport was the horse and carriage. Robert Anderson made the first crude EV in 1832 [5]. Just two years later, the first non-rechargeable battery-operated tricycle was introduced by Thomas Davenport. However, it took almost 1870 or later for EV to become a practical proposal. In 1874, David Salomons invented a lead-acid battery (that is, a rechargeable battery)-based EV. William Morrison from Des Moines, Iowa, created the first EV in the USA in the 1889–1891 period. This vehicle was almost an electrified wagon, yet it increased attention in EVs. With the advent of rechargeable batteries and less noisy EVs, such cars quickly became popular among urban residents at the end of the 19th century. At the turn of the century, many innovators took an interest in EVs because of the high level of market demand.

20th century – Major reforms in EV: The renowned inventor Thomas Edison regarded the EV as a worthier mode of transportation and researched the development of better batteries in the 19th century. In 1901, Ferdinand Porsche invented the Lohner Porsche Mixte, the initial hybrid vehicle powered by both gas engine and a battery. Following the introduction of the starter by Charles Keetering, the problem of manual cranking was removed, as it enabled electric ignition for hybrid vehicles. With the launch of the Model T with a starter, the EV market sales increased during the four-year period from 1908 to 1912 [6]. But due to the discovery of cheap Texas oil in the decade of the 1920s, the demand for EVs fell precipitously. After the 1970s, however, the prices of gasoline soared again, which created the way for introduction of the EV as a popular mode of transportation. Over the same period, the first manned electric-powered vehicle, named Lunar, was sent to the Moon by NASA, an event which helped to raise interest in EV. In 1973, General Motors (GM) created its first prototype Urban EV, which was displayed in a symposium on Low Pollution Power System Development. In 1979, due to the limitations of EV, such as their limited performance and range, the popularity of EVs began to decline again. In 1990, the California Air Resource Board (CARB) entrenched regulations that 2% of complete sold vehicle in California in 1998 ought to be zero emission vehicles (ZEVs), rising to 10% by 2003 [7]. In 1990–1992, new regulations were introduced by federal and state regulations which renewed the interest in EVs, resulting in the modifications to the EV structure to attain speed and performance comparable to that of gasoline-fuelled vehicles. In 1996, EV1 was release by GM which gained lot of interest in transport sector.

21st century – The acceptance of EVs into the global market: In the 21st century the carmaker Toyota released Prius onto the global market; this became an instant success. Then, in 2006, Tesla Motors announced the rollout of a luxury sports car with a range of more than 200 miles, which gave a boost to EV designs and battery technology [8]. Between 2009 and 2013, the US Energy Department infused in a nationwide charging infrastructure expansion. In 2010, GM announced the launch of the Chevy Volt, the first commercial Plug-in EV. In the same year, Nissan launched its first ZEV, the Nissan Leaf. With the rise in EV sales, research into the most expensive component of EVs, i.e. batteries, also accelerated. The dependency on oil and other fossil fuels was reduced by between 30% and –60% in the USA, which also reduced the carbon footprint of the transport sector by 20%. Today, consumers have a multitude of choice to opt for when buying an EV. The current situation holds a lot of potential for helping a nation's economy and growth in order to create a more sustainable future. Figure 2.1 shows brief remarkable inventions of EVs in 19th, 20th and 21st centuries.

2.1.2 Types of electric vehicles

The EVs are divided into three main categories, as shown in Figure 2.2. ICE-based vehicles are the most conventional type of vehicle, being based on the internal combustion engine drive line. Its operation depend mainly on petroleum- or diesel-based fuels which are non-renewable sources of energy and produce harmful emissions; therefore,, these are now being gradually replaced by battery-based vehicles known as electric vehicles (hereafter EVs). Figure 2.2(a) shows a conventional ICE-based vehicle.

Hybrid Electric Vehicle: These vehicles are equipped with ICE-based drive line and predominantly run on fossil fuels. The batteries are added in the drive train to ameliorate the efficiency of the whole vehicle

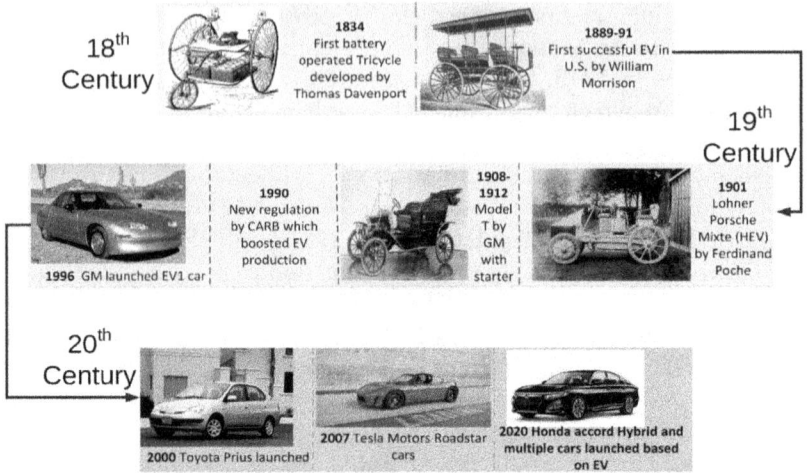

Figure 2.1 Brief history of EV launched in 18th, 19th and 20th century.

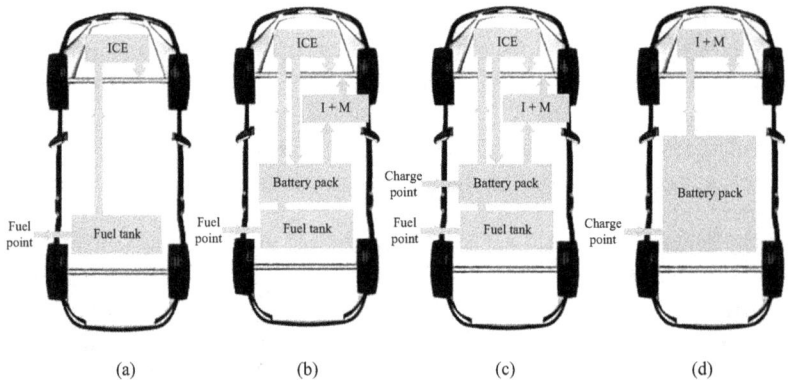

Figure 2.2 Type of vehicles (a) internal combustion engine (ICE) vehicle, (b) hybrid electric vehicle (HEV) (I-inverter and M-motor), (c) plug-in hybrid electric vehicle (PHEV) and (d) battery electric vehicle (BEV).

operation, as presented in Figure 2.2(b). In conventional vehicles, normally braking energy is lost as heat; by contrast, in HEV, this energy is referred to as regenerative braking energy and is used to recharge the batteries. These type of vehicles are unable to charge batteries from an external source (the grid). They have very low emissions, are easy to refuel and can be driven over long distances. Such vehicles are less powerful, heavier and have a complex structure as they combine a mechanical and an electric drive train. The Toyota Prius hybrid and the Honda Civic hybrid EVs are two example of HEVs which are

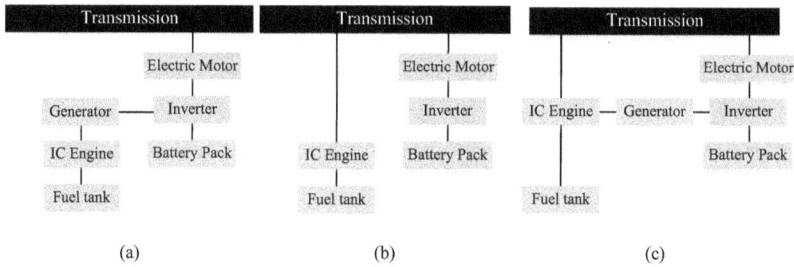

Figure 2.3 Types of HEV (a) series, (b) parallel and (c) series-parallel.

available on the Indian market. The HEVs are further classified into three types depending on the connection between the components as shown in Figure 2.3.

- Series HEV: In such vehicles, the ICE drives the electric generator which generates electric energy. This energy is utilized to charge the batteries packs and initiate the operation of motor as shown in Figure 2.3(a).
- Parallel HEV: In these type of vehicles, no separate generator is required. Parallel HEVs are propelled by both ICE and electric motors, as shown in Figure 2.3(b).
- Series-parallel HEV: Such vehicles combine series and parallel architecture. These vehicles can be started by either ICE or electric motor, as shown in Figure 2.3(c). The separate generator is also used.

Plug-in Hybrid Electric Vehicle (PHEV): These vehicles are akin to HEV, but have the freedom to charge batteries from the external source (grid), as shown in Figure 2.2(c). These vehicles are easy to refuel, having lower levels of fuel consumption and low emissions. Their price is higher and they are generally heavier vehicles. The Chevrolet Volt and Toyota Prius are among the most popular vehicles in this category. In the Indian market, the Volvo XC90 T8 and the BMW i8 are two of the models available in the PHEV category.

Battery Electric Vehicle (BEV): These vehicles are only equipped with an electric motor in drive line powered by batteries, as shown in Figure 2.2(d). These vehicles have no emission and their range is dependent on the type of batteries used. The problem associated with this type of vehicle is its battery price and capacity. The charging time depends on the charger configuration and its working power level. Such EVs have an advantage over other types of vehicle because of their simple architecture, operation and low maintenance. They have higher efficiency in comparison to HEV- and ICE-based vehicles. However, the lack of charging station and the short driving range limits its deployment in the Indian market. With advances in the fast-charging infrastructure and battery technology, it is likely that these shortcomings can be resolved in the near future. The Tesla and the Nissan Leaf are currently the world's best-selling EVs. The Tata Nexon EV, Mahindra eKUV100, Hyundai Kona Electric, and MG ZS EV are well represented in the Indian market.

2.1.3 The Basic Structure of EVs

A fundamental block diagram of the EV, comprising of the basic components, is shown in Figure 2.4. The major parts are the electric motor, the power converter, batteries and battery chargers.

Electric motors: In the EVs, electric motors are used to drive the vehicles. It is necessary that such motors should be able to deliver high torque at lower speeds and high power at higher speeds. In other words, EVs should operate at a wide speed range with constant torque and constant power. These motors should provide fast torque response with higher efficiency. For the better overall efficiency, it should also support high efficiency in regenerative braking. In addition, they should be reliable and robust.

In the earliest days, DC motors were used in EVs because they provided better and simple control when compared with AC motors. The DC motor would have best suited torque speed characteristic for traction. It required high maintenance due to the commutator, has lower

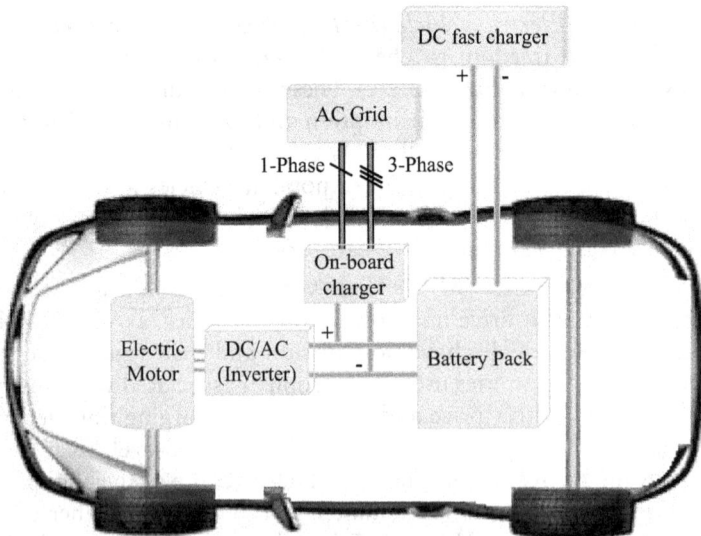

Figure 2.4 Basic structure of EV.

efficiency and is comparatively heavy. Today, because of technological advances, AC motors are also controlled in a similar manner. In some vehicles, such as the Renault Kangoo (2003) and the Renault ZOE (2011), a synchronous motor (SM) is used. A survey conducted over more than 200 EVs published in 2018 stated that the permanent magnet synchronous motor (PMSM) drive makes the highest contribution in the EVs, and HEVs drive trains [9]. PMSMs have higher power density, higher efficiency and lesser heating in comparison to IMs. These motors are also capable of providing higher torque at lower speed. The NdFeB type of PMSM is generally used because it provides high power density and high efficiency (97%) [10]. The Toyota Prius, Nissan Leaf and Chevy Volt are examples of vehicles which use interior types of PMSMs (IPMSMs). IMs are the second most used motors in EVs, as shown in Figure 2.5. One major point which can be noticed from the data depicted in Figures 2.5(a) and 2.5(b) is that the contribution of IM drives also increases with an increment in PMSM drives. The IMs are mainly preferred in high-powered vehicles (such as sports cars and trucks). The advanced IMs are used by Tesla in order to achieve higher efficiency and reduce reliance on rare earth magnets. Because of the restricted rare earth permanent magnets assets and related geopolitical factors, IMs will be used continuously in the EV and HEV sector. The permanent magnet-based motors are temperature-sensitive and may reduce the performance in the absence of effective cooling. However, IMs are very robust, require less maintenance, and are relatively inexpensive. The SRM are also used in EVs, generally suitable for constant power regions. The high EMI/EMC effect, low torque/power density, high torque ripples and acoustic noise are among the drawbacks of SRM drives [11]. The special converter topology is also required for controlling purposes. In a recent development, a Japan-based company,

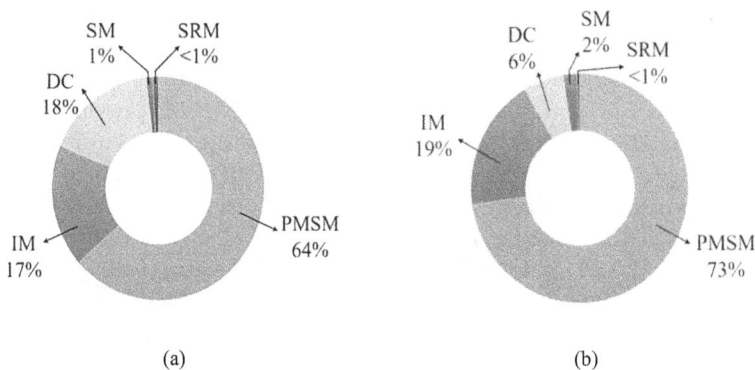

Figure 2.5 Percentage distribution of driving motors in EVs and HEVs (a) 1884–2016 years and (b) last 20 years [2].

Nidec, has developed a low-cost SRM for traction that exhibit performance close to that of IPMSM. Similarly, the Australian company Holden has developed the PHEV Holden ECOmmodore, which has a 50 kW SRM motor.

The multiphase motors (those with more than three phases) are under research for the on-road electric vehicles. These motors are already in use in applications such as electric traction and all electric aircraft [12]. The advantages provided by multiphase drives include reduced power/current rating per phase, reduced torque ripples, lower DC link capacitance, reduced stator copper losses, increased efficiency and fault tolerance [13].

Power converters: The semiconductor-based power converter is used to supply the electric motors of EVs. The various power semiconductor devices, such as BJT, GTO, MCT, IGBT and MOSFET, are available and suitable for EVs [14]. Of these, the GTO has a highest power handling capability (5000 V, 3000 A) of all of these devices, but it has a low switching frequency. The MOSFET are capable to operate at higher switching frequency, but have relatively low power handle capability (1000 V, 100 A). Perhaps the IGBT provides the best trade-off between power handling (1200 V, 400 A) and operating switching frequency. The power converters are designed to attain high power density and high efficiency. To operate the DC motor, DC-DC converters (choppers) are employed and AC motors are supplied by DC-AC converters (inverters). In the present trend, EVs are developed with an AC electric motor drive, which is controlled by an IGBT-based inverter through vector control (field-oriented control). The gallium nitride (GaN)- and silicon carbide (SiC)-based power semiconductor devices are the future technology. These have many advantages, such as faster switching speed, good thermal conductivity, smaller size, lower ON resistance and lower cost over silicon-based devices [15].

The electric motors of EVs operate as a variable speed drive controlled by power converters. The variable speed drive control techniques are generally classified into two parts: (1) scalar control; and (2) vector control. The scalar method, such as a V/f control, provides control over only the magnitude of the variables. Hence, its dynamic response is sluggish. Due to this performance limitation, vector control is introduced, which can vary both the phase and the magnitude of the variables. With advances in microcontrollers, field programmable gate array (FPGA) and DSP, the cost of implementing more advanced techniques is very much reduced. To achieve the better dynamic performance, vector control is implemented for the AC motors, where the decoupled control of torque and flux is possible akin to the DC motors. This decoupled control is commonly referred to as field oriented control (FOC).

An alternative to FOC was introduced in 1986, which provides the decoupled control of torque and flux by utilising nonlinear hysteresis controller and switching table [16]. In the Direct Torque Control (DTC) algorithm, both coordinate transformation and current control loops are not required, and they are hardly affected by parameter variations. The DTC scheme is simple to implement and it provides faster dynamic response with very small computational time [17]. However, DTC has a few disadvantages, including: it is difficult to regulate torque and flux at lower speed; it has high current and torque ripples; and it has variable switching frequency. The basic principle of DTC is to provide the regulation of only two variables (i.e., torque and flux). Hence, the inclusion of multiple constraints in the DTC scheme is no easy task. Therefore, the finite set model predictive control (FS-MPC) scheme for controlling motor torque, whuch is referred to as predictive torque control (PTC), has been introduced recently. It has many advantages over the conventional DTC algorithm. It provides better torque, and speed dynamic performance, lower torque ripple, lower average switching frequency and over current protection for the power converter [18]. In the FS-MPC scheme, nonlinearities and constraints can be included.

Batteries: The battery technology performs a non-trivial role in the EVs. Major aspects, such as driving range, recharge time and the cost of any EV, are dependent on its battery technology. The power density, energy density, cost, weight and volume are considered during the selection of batteries for the EVs. One more parameter, 'battery cycle life', is also taken into consideration as this provides the number of full charge and discharge cycles. After the completion of the battery life cycle, its nominal capacity falls below 60 to 80% and it becomes unusable. The type of batteries generally used in EVs is tabulated in Table 2.1. The NiCd batteries have longer life cycles, but they are costly and cadmium can cause environmental pollution if not disposed of properly. By contrast, NiMH batteries are harmless to the environment. Ni-based batteries have memory effect and therefore these can't be left charged for more than two days. Nowadays, lithium-ion batteries are dominant in the EV market and have low memory effect. They also achieve good performance at higher temperatures.

Battery charger: The basic classification this category for the EVs is presented in Figure 2.6 [20]. On the basis of power levels, battery chargers are categorized into three distinct types (Table 2.2). The chargers are divided into off-board and on-board on the basis of their mounting location. Off-board chargers is placed outside the EV while on-board chargers are placed inside the vehicle. Due to constraints of both size and weight, on-board chargers have restricted power levels and can be termed as either level 1 or level 2 chargers. The level 3 charger, by

Table 2.1 Type of batteries used in EVs

Type of battery	Cost	Life cycle	Power density (W/kg)	Energy density (Wh/kg)	Used in
Lead-acid	Low	200–300	120–200	30–40	Bertone Blitz (1992), GM EV1 (1996), Ford Ranger EV (1998) [19]
NiMH	Costly	300–500	250–1000	50–80	Citroën AX (1993), Toyota Prius (1997), Toyota Camry Hybrid (2007) [19]
Lithium-ion	Very costly	1000	1000–1500	100–150	BMW X5 (2003), Tesla Roadster (2008), Nissan Leaf (2010), Mahindra e-verito (2016), Tata Nexon (2020)

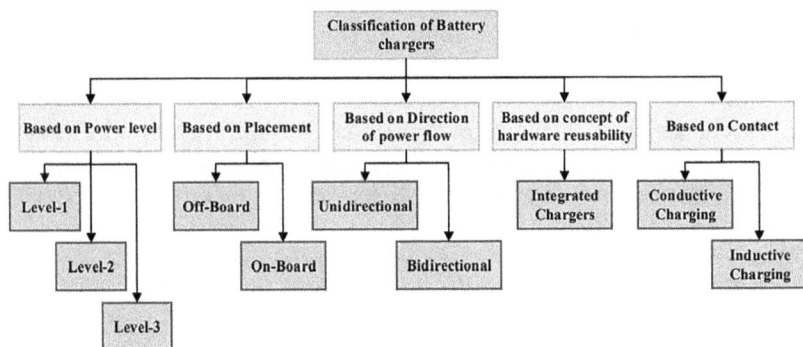

Figure 2.6 Classification of battery chargers.

Table 2.2 Classification of battery charger based on power levels (SAE-J1772)

Type of power levels	Charger placement	Expected power level (kW)	Charging time (hr)
Level 1	On-board (1-phase)	1.4–1.9	4–36
Level 2	On-board (1 or 3-phase)	4–19.2	1–6
Level 3	Off board (DC or 3-phase)	50–100	0.2–1

Table 2.3 Specifications of EVs available in global market

Manufacture/model	Type	Range (km)	Battery size (kW h)	Type of battery	Type of motor (kW)
Hyundai Kona Electric (2021)	EV	452	39.2	Li polymer	Front-PMSM
Nissan leaf SV plus (2021)	EV	346	62	Li-ion	PMSM
MG ZS EV (2021)	EV	419	44.5	Li-ion	PMSM
Tata Nexon EV (2020)	EV	312	30.2	Li-ion	Front-PMSM
Tesla model 3 (2017)	EV	568	82	Li-ion	Dual motor type (Front-IM Rear-IPMSM)
Mahindra e-verito (2016)	EV	140	21.2	Li-ion	Front-IM
Tesla Roadster (2008)	EV	354	53	Li-ion	IM

contrast, is classified as a fast charger. The integrated chargers are created by using the components of the existing drive lines to increase the power level of an on-board charger. So that size, weight and cost could save in EVs. In saved space, more battery packs can be accommodated to enhance the driving range of EVs. The bidirectional charger supports V2G functionality. In the V2G mode, the power available in the vehicle battery is again fed back to the grid. The process of charging EVs through the charging cables is knows as conductive charging. If the charging power is transferred to EVs through wireless or contactless medium, then they are categorised as inductive chargers or wireless chargers. Some specifications of EVs running in global market are given in Table 2.3.

2.1.4 EV market scenarios

Indian Scenario: The growth of vehicle ownership in India has increased rapidly, with ownership per 1000 people from 53 in 2001 to 167 in 2015 [21]. The Indian electric mobility market is enabled by policy, regulations, improved business models, investment opportunities and economics. In order to create demand for EV vehicles, the Department of Heavy Industry began incentives programmes, named FAME-I and FAME-II. Under the National Electric Mobility Mission Plan 2020 (NEMMP-2020), FAME-I gave support to 2.8 lakh EVs and HEVs with total demand incentives of INR 970 crore (USD 130 million),

India's EV financing market in 2030 Lakh Crore

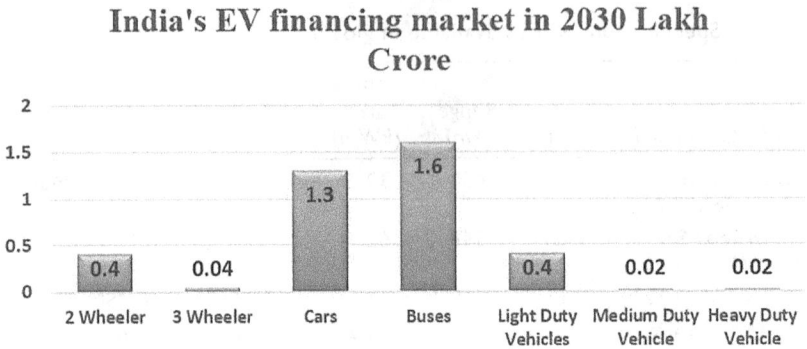

Figure 2.7 Size of India's EV financing market in 2030 (total INR 3.7 lakh crore).

a combined number which saved approximately seven crore litres of fuel and over seventeen crore kg of CO_2 emissions [22]. FAME-II started in 2019 with budget of INR 10,000 crore (USD 1.4 billion), aiming to drive large-scale EV adoption and the development of a charging infrastructure and EV ecosystem. Vehicles under FAME-II will eliminate seventy four Lakh tonnes of CO_2 emission over their lifetime. The economics of the EV market is improving gradually with time, driven by cost cutting in battery prices and improvements in vehicle operational efficiency. Now international firms are also investing in the Indian EV market. For example, the International Finance Corporation (IFC) has invested USD 8 million in Lithium Urban Technology based in Bengaluru in April 2018. By 2030, the capital cost of India's vehicles transition would be USD 266 billion, of which India's EV financing market will hold a share of USD 50 billion [23], as shown in Figure 2.7.

Global Scenario: The transport sector contributes 23% of GHG emission in EV countries. To eliminate this, the plug-in EV is introduced as an attempt to reduce GHG emission below 10% by 2050. China has become the world's major market for EVs, accounting for 35.4% of global EVs in 2017. Chinese consumers had bought 24.38 million passenger EVs in the previous year. BYD dominates the global EV market share with 13.2%, followed by Tesla with 9.9% and then other major global contributors [24]. In 2014, the percentage of passenger EV sales in different countries was as follows: Norway have 13.7%, Netherlands have 3.9%, Sweden have 1.5%, and the USA have 1.5%. Most other automobile markets have a share below 1%. Over the course of the past ten years, a 46–69% growth in the number of light EV is reported globally. (Of this amount, the top 5 EV models account for nearly 60% of total sales.) The global EV sales were about 2.1 million in 2018, a total increase of 64% compared to 2017. The contribution of EV sales by different countries in 2018 is 79% in the USA, 78%

Top 8 EV models sold in 2020

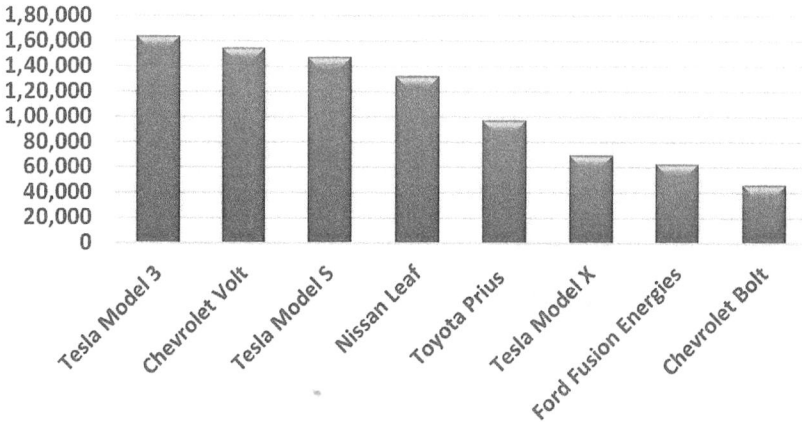

Figure 2.8 Top 8 EV models sales in the 2020.

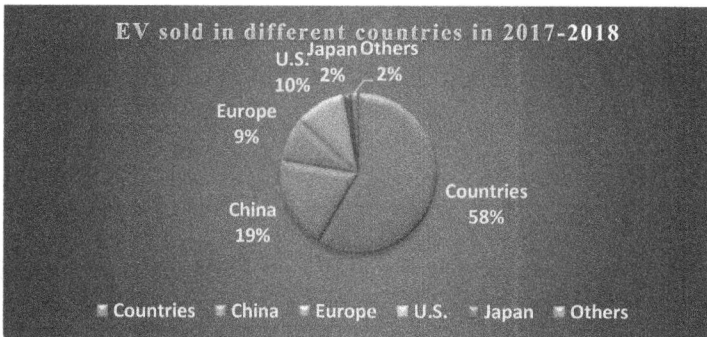

Figure 2.9 Number of EV sold in 2017–2018 in different countries globally.

in China and 34% in Europe. Figure 2.8 shows the chart of top 8 EV models sales in the 2020 [25]. Figure 2.9 shows the number of EV sold in 2017–2018 in different countries globally.

2.2 CURRENT TRENDS AND INNOVATIONS IN ELECTRIC VEHICLES

The research scope in EVs is vast. The major concern is how the driving range of EV can be increased. This objective is achieved by development and improvement in the battery technology and its charging system. In order to save space in EVs, research has been carried out to increase the energy and power density of the components (power electronics devices, batteries, electric motors) so that their size, weight and volume can be reduced. Some researchers have proposed the integrated solution to perform various operations through same drive line. The charging and propulsion operation of EV are not simultaneous processes. Therefore, te operation of the on-board charger is integrated with a bidirectional DC-DC converter in order to reduce the number of components in the EV [26, 27]. Similarly, traction inverter with traction motor stators windings are (as grid interface filter) used as fast bidirectional charger [28]. The advanced bidirectional on-board charger of EV supports V2G functionality [29].

2.2.1 Integration of EVs with a smart grid

The global aspects towards the adoption of EVs are increasing and the inclusion of EVs into the power grid mainly impact on load curves, transformer overloading and voltage unbalance. These issues are generally raised due to the unscheduled charging of larger numbers of EVs. The EVs are equipped with an energy storage system (batteries). Therefore, they are dealt in terms of distributed energy sources in a smart grid [30]. The V2G system can eliminate the power quality issues of the grid by voltage and frequency regulation, reactive power regulation, peak load shaving, valley filling, power factor correction and current harmonic filtering [31]. The V2G system also provides the backup to renewable energy sources. Along with V2G, vehicle to home (V2H) or vehicle to building (V2B) can also be an option [32]. In the V2G operation, battery life cycles are reduced, but they provide the option to earn incentives by injecting the available battery power back into the grid.

The block diagram of EV as presented in Figure 2.3 shows that for levels 1 and 2, the on-board charger is used to charge the vehicle. In level 2, the vehicle can be charged through three-phase connection. This charger converts the available AC from the grid into DC in order to charge the battery through the basic three-phase IGBT-based converter as shown in Figure 2.10.

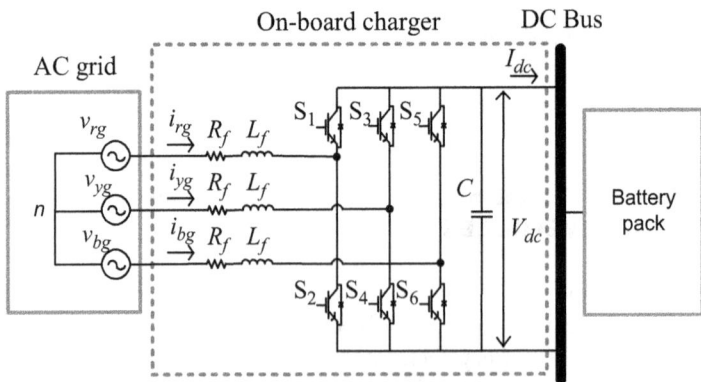

Figure 2.10 Three-phase voltage source converter of on-board charger.

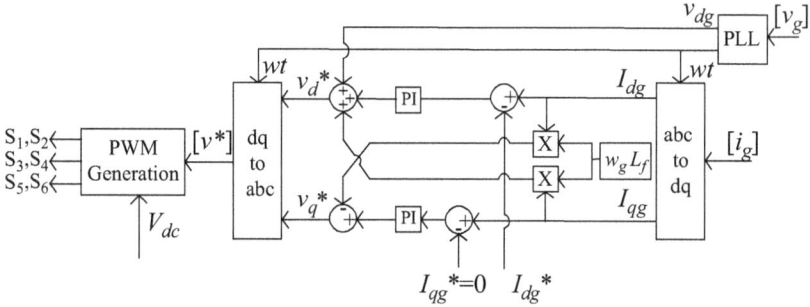

Figure 2.11 Single synchronous reference frame algorithm.

This converter should be bidirectional so that power can flow in both directions. The power flow is controlled by using the synchronous reference frame algorithm, as shown in Figure 2.11 [33]. The Phase Lock Loop (PLL) is used to evaluate the grid position. By using this, grid voltage and current are transformed to synchronous reference frame (dq). These time-varying quantities (v_g, i_g) are DC in nature when converted to synchronous reference frame. These DC quantities are easily regulated to their reference values by using the PI controller as shown in Figure 2.11.

The active and reactive powers are defined by:

$$P(t) = 1.5 \left[v_{dg}(t) i_{dg}(t) + v_{qg}(t) i_{qg}(t) \right] \tag{2.1}$$

$$Q(t) = 1.5 \left[-v_{dg}(t) i_{qg}(t) + v_{qg}(t) i_{dg}(t) \right] \tag{2.2}$$

In the balanced grid condition, v_{dg} is equal to the peak value of grid voltage and v_{qg} is equal to zero. Therefore, I_{dg}^* is calculated through (Equation 2.1). The value of I_{qg}^* is set to zero for obtaining unity power factor.

The simulation is performed by considering grid voltage of 415 V, 50 Hz. The DC bus voltage is assumed at 650 V. For the charging mode, G2V results are obtained by setting I_{dg}^* to 4 A as shown in Figure 2.12. For V2G, I_{dg}^* is set to -5A. It observed that grid voltage and grid current are in phase. Therefore, power flow is now reversed. The results for V2G are shown in Figure 2.13. The single synchronous reference frame algorithm is only suitable when the grid voltage is balanced. The random charging of a large number of EVs may destabilize the grid and generate unbalanced in-grid voltage. In that scenario, a single synchronous reference frame control algorithm is not capable to provide the desired performance and also pollute the grid with odd order harmonics [34,35]. The double synchronous reference-based control algorithm is used to mitigate this problem. In this control, positive $\left(I_{dg}^+, I_{qg}^+ \right)$ and negative sequence components $\left(I_{dg}^-, I_{qg}^- \right)$ are regulated to their reference values. The function of sequential extraction block is to extract the positive and

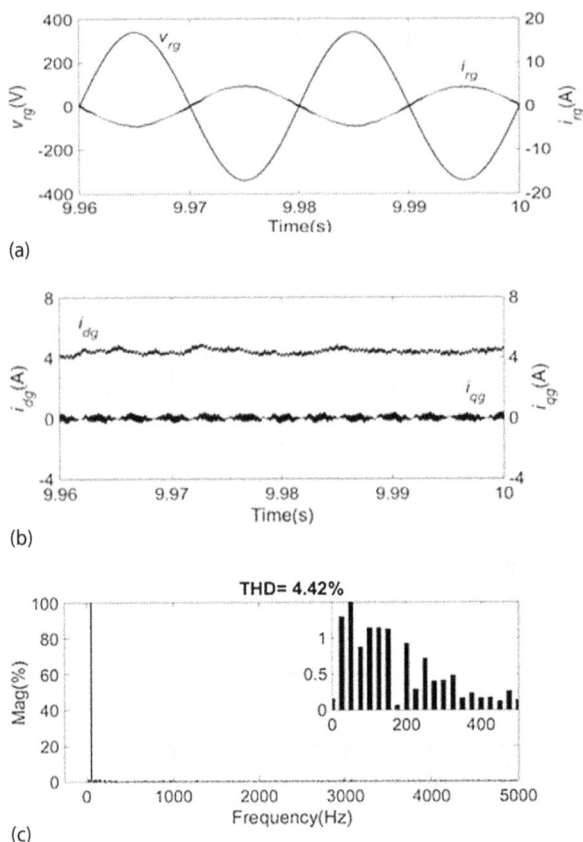

Figure 2.12 (a) Grid voltage and grid current, (b) dq-components of grid current, (c) Grid current THD during G2V operation.

negative sequence components of the grid current. The principle of extraction block is based on the decoupled network as explained in [36]. The block diagram of control algorithm is presented in Figure 2.14.

The control algorithm is simulated on MATLAB with the consideration of unbalanced voltage as presented in Figure 2.15. Comparative results are presented in Figure 2.16.

It is observed from the grid current THD that due to unbalanced grid voltage, third harmonics are injecting into the grid when controlled with a single synchronous reference frame algorithm. These harmonics are reduced by using the double synchronous reference frame control. The I_{dg}^{+} is regulated to the reference value of 8A and remaining components $\left(I_{qg}^{+}, I_{dg}^{-}, I_{qg}^{-}\right)$ are regulated to zero reference as shown in Figure 2.16(e). It is observed that grid current THD (3.55%) is maintained under the allowable limits with unbalanced grid voltages.

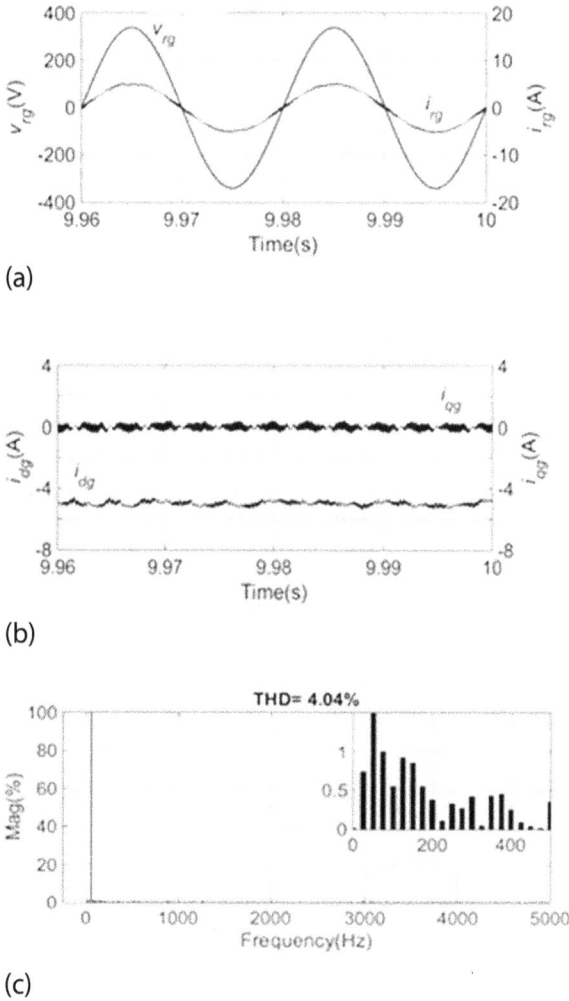

Figure 2.13 (a) Grid voltage and grid current, (b) dq-components of grid current, (c) Grid current THD during V2G operation.

The impacts due to unscheduled charging/discharging of the EVs on the power grid can be resolved by smart optimization schemes. Various optimized techniques are developed to achieve the objective. The particle swarm optimization is one of the intelligent technique for optimal scheduling. The energy resources (EVs) management methodology by considering the cost and load factor to provide the prior scheduling of EVs is presented in [37]. The model predictive control-based algorithm is developed for scheduling the charging to reduce the impact on the power grid [38].

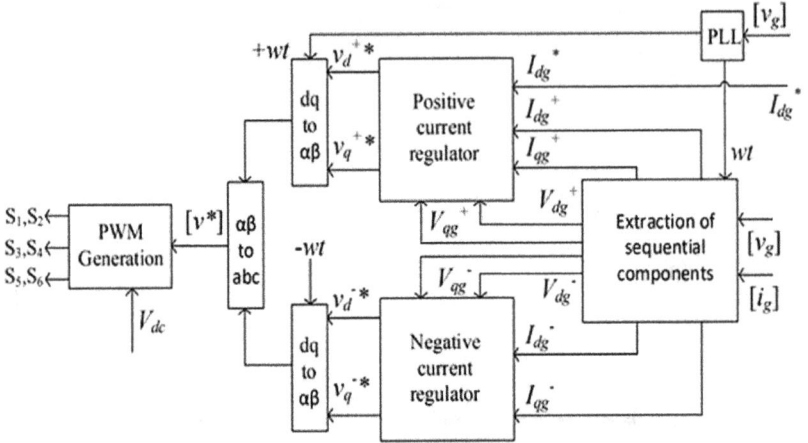

Figure 2.14 Double synchronous reference frame based control algorithm for unbalanced grid voltage condition.

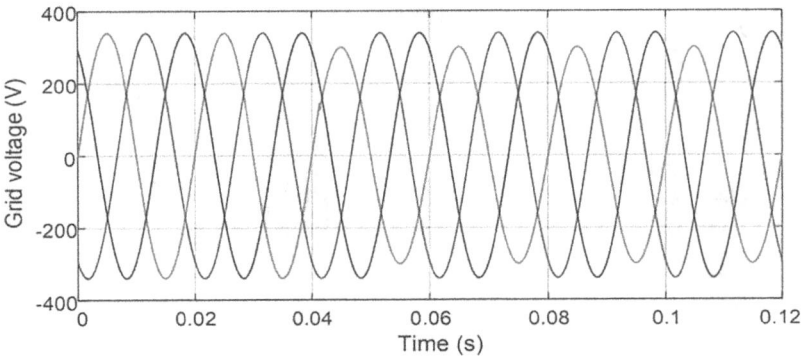

Figure 2.15 Grid voltage.

2.2.2 Infrastructure and technology implementation

EV implementation as an energy buffer can create more problems in the Distribution Network (DN). Several researches have pointed out that the large-scale implementation of EVs in charging mode can have a negative impact on DN. It is possible that the power demand can increase significantly in peak hours and may cause a drop in voltage. To tackle such issues scheduled charging and discharging is very essential. EVs can be integrated into the power distribution network in several ways, for example as dynamic loads which draw electricity from the grid during charging (G2V) and the Energy Storage System (ESS), which feeds electricity into the grid the in case of scarcity of power (V2G). However, the mobility (spatial-location) of EVs and low individual storage capacities makes it unrealizable for V2G

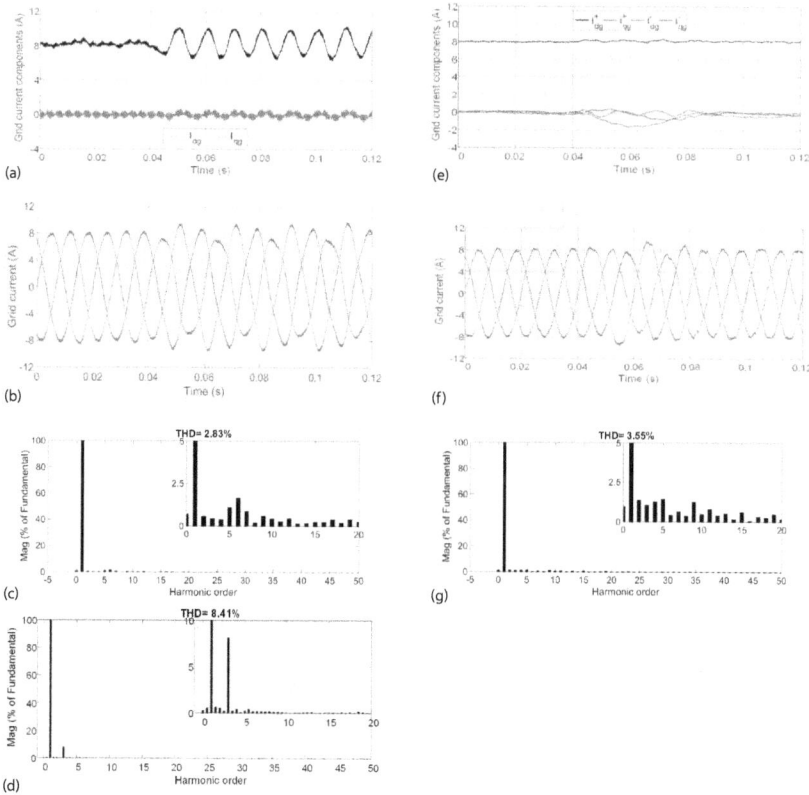

Figure 2.16 With single synchronous reference frame (a) dq-components of grid current, (b) grid current, (c) grid current THD under balanced grid voltage, (d) grid current THD under unbalanced grid voltage and With double synchronous reference frame (e) positive and negative sequence components of grid current, (f) grid current, (g) grid current THD.

operation. In order to solve this problem, large-scale EVs are clustered in various ways using control strategies and optimization algorithms to fulfil utility- and consumer-oriented objectives during V2G services. This aggregation of EV converts into a unit-controllable distributed energy source that can be integrated with the energy market to support regulation and management of smart grid [39]. Another issue is the extra revenue expenditure for the protection and metering system. These expenses nullify the V2G benefit acquired by bidirectional power flow. Further studies are also in process to explore the feasibility of unidirectional power flow for low-capacity V2G transactions. The concept of VPP facilitates the aggregation scenario in which control and information exchange is easily established between utility and EV fleets. Depending on the control strategy and aggregation type,

VPP framework models are of the following types: centralized, hierarchical and distributed control. In the centralized control model, the data exchange and decision-making is carried out by VPP central control centre (VPPCC). In the distributed control model, the decision and data flow is managed in a distributed manner. By contrast, hierarchical control uses a spatial VPP model for data exchange and decision-making [40].

The EV infrastructure consists of a smart meter as a key component to enable pricing, energy management, measurement, communication and control in real time. To make the EV charge schedule smarter, the optimization algorithms are utilized to meet both utility- and consumer-oriented goals. This is accomplished using advance bidirectional data exchange over the communication network in the smart grid context. Therefore, three major components of EV infrastructure in the V2G scenario are: (i) EV smart scheduling; (ii) smart metering; and (iii) communication and control network.

(i) EV smart scheduling – EV unscheduled and uncontrolled charging and discharging may cause the overload of the power distribution network and increased demand in peak hours which can cause reduction in the efficiency of the supply system. In order to resolve this issue, several technique have been introduced under smart charging schemes that can pursue different objectives. To schedule the EVs in an optimized manner, an intelligent algorithm is required to benefit the EV market. The objectives may vary for different EV systems, including charging cost economy, GHG emission, battery life, losses, and other matters. In [41], modified PSO is used for the smart scheduling of EV. In a few research studies, EV are controlled to fulfil demand response programme objectives. An optimized price algorithm is discussed in [42] to manage EV charging. To facilitate V2G operation, RFID technology is used. In [43], a method is presented to reduce power losses and enhance voltage profile in a smart grid for the V2G environment.

(ii) Smart metering (SM) – To analyse report energy usage and operate energy management system in the smart grid, smart meters are implemented widely. It is a core component which provide real-time information of demand and usage. SM does the task of energy forecast and energy pricing more effectively [44]. The Advanced Metering Infrastructure (AMI) constitutes a Meter Data Management System (MDMS), a Home Area Network (HAN), SMs, computer system, software, sensor network and communication technologies. The communication network in the AMI model can be wireless, PLC-based or any other network to provide two-way communication. The information collected by AMI can be used to implement decision-making and control algorithms. So, it can be said that EV smart scheduling is not practically possible without smart metering infrastructure.

(iii) Communication technology – The bidirectional information flow is the basic requisite of a smart grid network to enable the demand response programme which can control large-scale deployment of DES over widespread geographical locations. In this era of wireless communication, the technology can provide a feasible alternative for economic and large-scale coverage in the V2G scenario. Depending on the V2G interaction, the communication technology-based solutions can be envisioned in two ways. First, information flow from sensor network to SM and second from SM to grid operator data centre. The former can be done using wireless or PLC communication while the later can be established using the 3G mobile network, WiMax and 4G LTE. The challenges in terms of communication technology deployment is its reliability, cost and short range, resulting in packets delay and a falling success ratio. On the other hand, cybersecurity for EVs and power markets should be assured to prevent cyber-attacks such as price tampering, data manipulation, malicious software and system congestion. It is essential to ensure the security of EV services at visiting networks [45].

Figure 2.17 shows the major parts and networking of a V2G system. This system comprises of six major components:

  (i) Energy sources and utility
 (ii) Independent operators and aggregators
(iii) Charging infrastructure components
 (iv) Bidirectional electrical energy and information flow
  (v) Metering and control system
 (vi) Battery charger and management

2.2.3 Impacts and impediments of EV penetration

Considering the large-scale deployment of EVs, DES that can be harnessed by V2G interaction has several impressions over the grid operation. In addition, they can act as quick-response type loads for the distribution network to participate in a demand response programme. There are different impediments and barriers in V2G implementation, including battery degradation, modifications in the existing infrastructure, bidirectional communication network requirement, impact on distribution network parameters, losses, and other technical problems.

• Battery degradation – This depends on several factors, such as the amount or rate of power drawn, DoD and cycling frequency. Prediction of battery depletion is difficult because the technologies are changing continuously. However, one parameter can help in the estimation of the life cycle, i.e. equivalent series resistance [46].

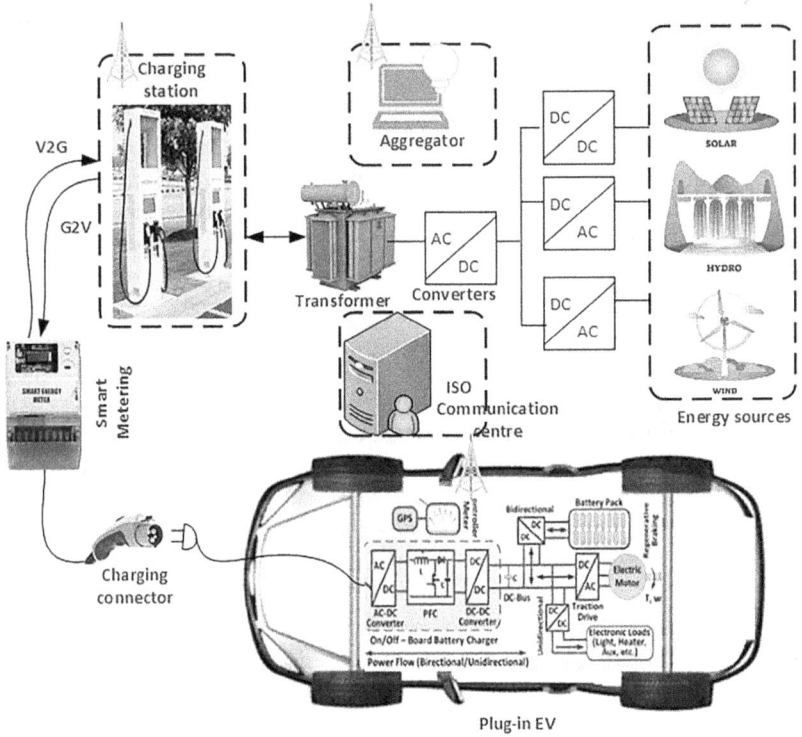

Figure 2.17 Components of V2G system.

- Impact on the existing distribution network – If the charging and discharging operation take place for a large number of EVs simultaneously, it can create overloading and the transformer can be at a dangerous level as a result of such instances. Likewise, the main feeder conductor would also be overloaded. Power loss in charging and discharging is one of the most prominent economic concerns for a distribution operator. The following impact evaluation is required in the case of V2G interactions:

  (i)   Impact on the distribution transformer
  (ii) Conductor/cables loading
  (iii) Overloading
  (iv) Voltage deviations
  (v)   Power loss
  (vi) Stability and reliability of system
  (vii) Power quality

- Impact of EV penetration level on the grid – The need for green energy has raised the popularity of EVs. It is vital to access the status of grid, whether or not it is capable of handling large EV penetration.

However, in the current context, the grid is dealing with several issues, such as charging procedures, random power demand management and stability problems in the V2G scenario. So, it can be inferred that the grid is not currently fully prepared for large-scale EV penetration. In addition, unscheduled EV charging may have an influence on economic cost and emissions and also an impact on the distribution network [47].

2.3 STANDARDS AND POLICIES ADOPTED BY DIFFERENT COUNTRIES

2.3.1 Standards and codes for EV implementation

Across the globe, various organisations, including IEEE, SAE, ISO, IEC, and so on are working to establish universal standards. These standards ensure the safety of the user and reliability, and the performance of the component or device. The SAE-J1772 standard defines the charging levels for conductive charging. This standard categorized the charging levels into level 1, level 2 and level 3, according to the power levels as shown in Table 2.2.

The level 3 chargers are known as fast charger and capable for charging batteries within 0.2 to 1hr. The vehicles are equipped with charging port so that they can be connected to a charging outlet. The combined charging system (CCS) allows the AC and DC charging through the same charging port while CHArge de MOve (CHAdeMO) (the Japanese conductive charging standard) has a separate port for AC charging. The IEC 1000-3-2, SAE-J2894 and IEEE 1547, and standards presents the allowable harmonics and DC current injection into the grid. These power electronics-based components have power switches; their switching operation generates dv/dt and di/dt and their performance degrades due to electromagnetic disturbance. This is known as electromagnetic interference (EMI). Therefore, these components should be compatible with electromagnetic compatibility (EMC) standards [48]. Among the most commonly used international standards in EVs are those cited in Table 2.4. The electric components used in EVs should be compatible to ingress protection (IP) standards so that there is no accumulation of dust and water. For example, IP6k9k provides protection against the powerful high pressure and high temperature water jets.

The ARAI (Automotive Research Association of India) regulates or publishes the automotive industry standards (AIS) in India on behalf of the Automotive Industry Standards Committee (AISC). The commonly used standards developed by ARAI are listed in Table 2.5. The ARAI also provides the certification and testing facility of the EVs. The Bureau of India Standards (BIS) also published standards (IS codes) for the EVs mainly related to the charging system or electric vehicle supply equipment (EVSE).

Table 2.4 International standards for EVs

Standard	Purpose
SAE-J2894	Allowable harmonics and dc current injection
IEEE 1547	
IEC 61000	EMC
IEC 60529	Ingress protection (IP): Provide protection against intrusion, accidental contact, dust and water.
SAE-J1772	Charging levels for conduction charging
SAE-J1773	For inductive charging
SAE-J2954	
SAE-J1634	Standard test for determination of battery energy consumption and range
SAE-J1715	Definitions for EV and HEV configurations
SAE-J2344	Electric vehicle safety guidelines
SAE-J3072	Requirements for interconnection of grid support inverter (on-board)
ISO 6469-1	Safety requirement for rechargeable energy storage system (RESS)
ISO 6469-2	Operational safety specifications
ISO 6469-3	Safety requirement for vehicle
ISO 6469-4	Safety requirement for post-crash condition
ISO 8714	Test procedure for calculating energy consumption and range
ISO 15118	Vehicle to grid (V2G) communication interface
ISO 21782	Test specification for electric traction components

2.3.2 Schemes introduced by the Indian government

The FAME (Faster adoption and manufacturing of Electric and Hybrid vehicles) scheme was launched by the Indian government in 2015 for the two years. Under this scheme, around INR 359 Cr is distributed for 20.8 lakh vehicles. This scheme was later further extended to September 2018. Phase two of FAME was started from 1 April 2019 for the three years and government approved the INR 10,000 Cr amount. The objective of this phase to support 7090 electric busses, 55,000 electric four-wheelers, five lakh electric three-wheelers and ten lakh electric two-wheelers [49]. A total of 74,527 vehicles have been sold, as a result is estimated that some 1,97,74,274 litres of fuel will be saved and CO_2 emissions will be reduced by around 4,50,03,341 kg. The Indian government has reduced the goods and service tax (GST) rate from 12% to 5% on EVs. The GST rate on charging stations and chargers is also reduced, from 18% to 5%. Along with the central, several states (Delhi, Maharashtra, Uttar Pradesh, Gujarat, and Karnataka) in India also implemented or introduced EV policies to create the market and support the manufacturing companies. The Delhi Electric Vehicle policy (2020) aims to provide incentives on EV purchasing [50].

Table 2.5 Indian standards for EVs

Standard	Purpose
AIS 004(3)	EMC
AIS-038	Constructional and functional safety requirements such as electric shock, direct contact, indirect contact, water effects
AIS-039	Measurement of electrical energy consumption (Wh/km)
AIS-040	Method for measuring the vehicle range
AIS-048	Safety requirement for traction batteries
AIS-138(1)	Conductive AC charging system
AIS-138(2)	Conductive DC charging system
IS 16894	Safety requirement for secondary batteries and battery installation
IS 13514	Performance and endurance test for secondary batteries (except lithium-ion)
IS 17387	Battery management system related general safety and performance
IS 17017	Conductive charging system requirements
IS 17017:Part21-Sec1	EMC requirement for On-board charger
IS 17017:Part21-Sec2	EMC requirement for Off-board charger

The Uttar Pradesh state's EV Manufacturing and Mobility Policy 2019 focuses on increasing the use of HEVs and EVs. The objective is to install 2 lakh charging stations and 1 million EVs (a 70% share of EVs in public transport) by 2030 [51].

2.4 FUTURE AREAS OF RESEARCH IN EV IMPLEMENTATION

In order to achieve goals such energy security, better air quality, reduced noise pollution and a reduction in the emission of greenhouse gases, an emphasis on the adoption and promotion of EVs can be a promising measure. However, there are still a number of challenges that need to be addressed. Firstly, EVs have limited driving ranges and those utilizing a single energy source are unable to deliver better dynamic performance. Integrating more than one energy source in an EV drivetrain may increase its dynamic performance, yet range anxiety remains a major concern among EV owners. It can only be resolved by the establishment of an efficient and widespread network of charging infrastructure.

Another challenge is the proper disposal of the batteries and fuel cells stacks which are being used as energy resources in EVs. These batteries contains dangerous heavy metals which can have hazardous implications when disposed of inappropriately and may create threats for the environment and public health. With an increase in renewable energy system (RES) generation, the objective of EVs interacting with RESs will provide different future aspects for research in EV implementation. In the case of large-scale RES integration with EV, problems related to the disparity between demand and supply, reactive power injection, and reduced frequency control all need to be addressed. In this domain, few research studies have been published which discuss the impact of large-scale RESs penetration integrating with EVs [52, 53].

2.4.1 Research for new energy storage technologies and infrastructure

The RES-generating electricity will charge EV batteries. In the case of supply shortages, the EV battery can discharge by switching to V2G mode. Most of the EVs utilize Li-ion batteries in which power capacity is affected by an increase in charging and discharging cycles. Thus, battery capacity fades out quickly. This creates a concern in vehicle owners over whether or not to participate in the DSM programme. In addition, Li-ion batteries has considerable potential for improvement. There are other options which can help in the range extension of EVs, e.g. Li-air batteries. Still there is a huge scope for improvement in battery technology, which can benefit consumers by alleviating range anxiety and safe operation. The successful deployment of EV in any future scenario requires the following [54]:

- Deployment of an efficient charging infrastructure.
- Safety, reliability, durability and compatibility of chargers available at charging stations.
- Incentive-based charging cost.
- Suitability for V2G operation with smart meters and communication networks.
- Charging schedules and setting the time limits and rules.
- Standardization around the EV charging station.

2.4.2 Feasibility of V2G support to renewable energy sources

There are a number of major challenges which need to be addressed and resolved for the successful implementation of V2G interaction. The first and foremost requisite for V2G integration is that EV should be consisting of bidirectional power converters and advanced communication devices. Secondly, the charging infrastructure should be outfitted with smart charging points that can send and receive information to the control unit of EVs. Third, to manage the control strategy of a grid operator for the vehicle

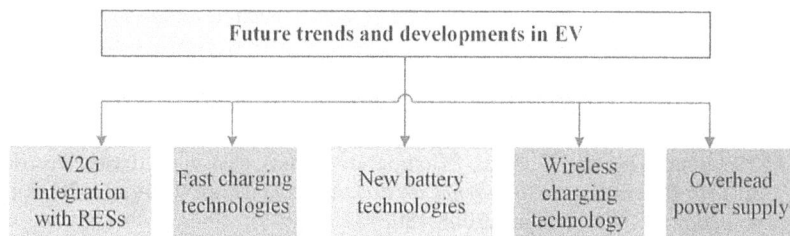

Figure 2.18 Future trends and development in EV implementation.

participating in V2G integration. Fourth, EVs are mostly treated as energy source or energy storage. Since, it is mostly on the fleet so its distribution cannot be anticipated in advance. This can create congestion issues while scheduling the energy flow in a power distribution network. The e-mobility pilot project Zem2All inaugurated in Malaga city, Spain in April 2013 is an example of a V2G project. It has 23 CHAdeMO DC fast-charging points and 6 bidirectional chargers [55]. Research and pilot projects are already established to enhance the efficiency and lower the cost of the EV DC fast-charging infrastructure.

With V2G proliferation, other systems required to establishing a sustainable EV scenario becomes essential, such as charge scheduling, a Virtual Power Plant (VPP) and smart metering. Other future trending topics which can help in better EV adoption are shown in Figure 2.18.

REFERENCES

[1] "Air pollution-transportation linkage," California Air Resources Board Office of Strategic Planning, 1989.

[2] Online: https://niti.gov.in/writereaddata/files/document_publication/EV_report.pdf (NITI A).

[3] Gupta, Dipti, Frédéric Ghersi, Saritha S. Vishwanathan, and Amit Garg. "Achieving sustainable development in India along low carbon pathways: Macroeconomic assessment," *World Development*, vol. 123 (2019): 104623.

[4] Larminie, James, and John Lowry. *Electric vehicle technology explained.* Hoboken, NJ: John Wiley & Sons, 2012.

[5] Holms, Ann, and Rony Argueta. "A technical research report: The electric vehicle," *Argueta–6-7*, March 11, 2010.

[6] Brooke, Lindsay. *Ford model T: The car that put the world on wheels.* Minneapolis, MN: Motorbooks, 2008.

[7] Collantes, Gustavo, and Daniel Sperling. "The origin of California's zero emission vehicle mandate," *Transportation Research Part A: Policy and Practice*, vol. 42, no. 10 (2008): 1302–1313.

[8] Thomas, V. J., and Elicia Maine. "Market entry strategies for electric vehicle start-ups in the automotive industry–Lessons from Tesla Motors," *Journal of Cleaner Production*, vol. 235 (2019): 653–663.

[9] Bazzi, A. M., Y. Liu, and D. S. Fay. "Electric machines and energy storage: Over a century of technologies in electric and hybrid electric vehicles," *IEEE Electrification Magazine*, vol. 6, no. 3 (September 2018): 49–53.

[10] Miller, T. J. E. *Brushless Permanent-Magnet and Reluctance Motor Drives*. London, UK: Oxford University Press, 1989.

[11] Hashernnia N., and B. Asaei. "Comparative study of using different electric," In *18th International Conference on Electrical Machines Vilamoura*, 2008, pp. 1–5, doi:10.1109/icelmach.2008.4800157

[12] Salem, A., and M. Narimani. "A review on multiphase drives for automotive traction applications," *IEEE Transactions on Transportation Electrification*, 2019, pp. 1–1.

[13] Levi, E., R. Bojoi, F. Profumo, H. A. Toliyat, and S. Williamson. "Multiphase induction motor drives – A technology status review," *IET Electric Power Applications*, vol. 1, no. 4 (July 2007): 489–516.

[14] M. Fracchia, T. Ghiara, and M. Marchesoni. "Generalized design of power converters for electric vehicle," in *Proc. Int. Electric Vehicle Symp.*, 1992, no. 16.01.

[15] Rajashekara, K. "Present status and future trends in electric vehicle propulsion technologies," *IEEE Journal of Emerging and Selected Topics in Power Electronics*, vol. 1, no. 1 (March 2013): 3–10, doi:10.1109/JESTPE.2013.2259614.

[16] Takahashi, I., and T. Noguchi. "A new quick-response and high-efficiency control strategy of an induction motor," *IEEE Transactions on Industry Applications*, vol. IA-22, no. 5 (September 1986): 820–827.

[17] Sharma, S., M. Aware, A. Bhowate, and E. Levi. "Performance improvement in six-phase symmetrical induction motor by using synthetic voltage vector based direct torque control," *IET Electric Power Applications*, vol. 13, no. 11 (November 2019): 1638–1646.

[18] Bhowate, A., M. V. Aware, and S. Sharma. "Predictive torque control algorithm for a five-phase induction motor drive for reduced torque ripple with switching frequency control," *IEEE Transactions on Power Electronics*, vol. 35, no. 7 (July 2020): 7282–7294, doi:10.1109/TPEL.2019.2954991

[19] de Santiago, H. Bernhoff, B. Ekergård, S. Eriksson, S. Ferhatovic, R. Waters, and M. Leijon. "Electrical motor drivelines in commercial all electric vehicles: A review," *IEEE Transactions on Vehicular Technology*, vol. 61 (Feb. 2012): 475–484.

[20] Yilmaz, M. and P. T. Krein. "Review of battery charger topologies, charging power levels, and infrastructure for plug-in electric and hybrid vehicles," *IEEE Transactions on Power Electronics*, vol. 28, no. 5 (May 2013): 2151–2169.

[21] Online: https://community.data.gov.in/registered-motor-vehicles-per-1000-population-from-2001-to-2015/

[22] Department of Heavy Industry. https://fame2.heavyindustry.gov.in/

[23] Inc42 Media. "Electric Vehicle Market Outlook Report 2020," 2020. https://inc42.com/reports/electricvehicle-market-outlook-report-2020/

[24] EV-Volumes—The Electric Vehicle World Sales Database. Available online: http://www.ev-volumes.com/country/total-world-plug-in-vehicle-volumes/

[25] Wolfram, Paul, and Nic Lutsey. "Electric vehicles: Literature review of technology costs and carbon emissions," In *The International Council on Clean Transportation: Washington, DC, USA*, 2016, pp. 1–23.

[26] Lee, Y., A. Khaligh, and A. Emadi. "Advanced integrated bidirectional AC/DC and DC/DC converter for plug-in hybrid electric vehicles," *IEEE Transactions on Vehicular Technology*, vol. 58, no. 8 (Oct. 2009): 3970–3980, doi:10.1109/TVT.2009.2028070

[27] Dusmez, S., and A. Khaligh. "A compact and integrated multifunctional power electronic interface for plug-in electric vehicles," *IEEE Transactions on Power Electronics*, vol. 28, no. 12 (Dec. 2013): 5690–5701, doi:10.1109/TPEL.2012.2233763

[28] Sharma, S., M. V. Aware, and A. Bhowate. "Integrated battery charger for EV by using three-phase induction motor stator windings as filter," *IEEE Transactions on Transportation Electrification*, vol. 6, no. 1 (March 2020): 83–94, doi:10.1109/TTE.2020.2972765

[29] Mukherjee, J. C., and A. Gupta. "A review of charge scheduling of electric vehicles in smart grid," *IEEE Systems Journal*, vol. 9, no. 4 (Dec. 2015): 1541–1553, doi:10.1109/JSYST.2014.2356559

[30] Sioshansi, R., and P. Denkolm. "The value of plug-in hybrid electric vehicles as grid resources," *Energy J.*, vol. 31, no. 3 (2010): 1e23.

[31] Breucker, S. D., P. Jacqmaer, D. Brabandere, J. Driesen, and R. Belmans. "Grid power quality improvements using grid-coupled hybrid electric vehicles," In *Proc. Power Electron., Mach. Drives Conf.*, 2006, pp. 505–509.

[32] Shemami, M.S., M.S. Alam, and M.S.J. Asghar. "Reliable residential backup power control system through home to plug-in electric vehicle (H2V)," *Technology and Economics of Smart Grids and Sustainable Energy*, vol. 3, no. 1 (Dec. 2018): 8.

[33] Sharma, S., M. Aware, Y. Tatte, J. K. Pandit, and A. Bhowate. "A split three phase induction motor for battery charging application," In *2016 IEEE International Conference on Power Electronics, Drives and Energy Systems (PEDES)*, 2016, pp. 1–6, doi:10.1109/PEDES.2016.7914515

[34] Sharma, S., M. Aware, and A. Bhowate. "Control algorithm for G2V/V2G operation under unbalanced grid condition," In *2017 7th International Conference on Power Systems (ICPS)*, Pune, 2017, pp. 188–193.

[35] Suh, Y., Y. Go, and D. Rho. "A Comparative study on control algorithm for active front-end rectifier of large motor drives under unbalanced input," *IEEE Transactions on Industry Applications*, vol. 47, no. 3 (May–June 2011): 1419–1431.

[36] Rodriguez, P., J. Pou, J. Bergas, J. I. Candela, R. P. Burgos, and D. Boroyevich. "Decoupled double synchronous reference frame PLL for power converters control," *IEEE Transactions on Power Electronics*, vol. 22, no. 2 (March 2007): 584–592.

[37] Morais, Hugo, Tiago Sousa, Zita Vale, and Pedro Faria. "Evaluation of the electric vehicle impact in the power demand curve in a smart grid environment," *Energy Conversion and Management*, vol. 82 (2014): 268–282, ISSN 0196-8904, doi:10.1016/j.enconman.2014.03.032

[38] Shi, Y., H. D. Tuan, A. V. Savkin, T. Q. Duong, and H. V. Poor. "Model predictive control for smart grids with multiple electric-vehicle charging stations," *IEEE Transactions on Smart Grid*, vol. 10, no. 2 (March 2019): 2127–2136, doi:10.1109/TSG.2017.2789333

[39] Jaiswal, S., and M. S. Ballal. "Optimal load management of plug-in electric vehicles with demand side management in vehicle to grid application," *2017 IEEE Transportation Electrification Conference (ITEC-India)*, 2017, pp. 1–5, doi:10.1109/ITEC-India.2017.8356942

[40] Raab, A. F., M. Ferdowsi, E. Karfopoulos, I. Grau Unda, S. Skarvelis-Kazakos, Panagiotis Papadopoulos, E. Abbasi et al. "Virtual power plant control concepts with electric vehicles," In *2011 16th International Conference on Intelligent System Applications to Power Systems*, IEEE, 2011, pp. 1–6.

[41] Soares, João, Hugo Morais, Tiago Sousa, Zita Vale, and Pedro Faria. "Day-ahead resource scheduling including demand response for electric vehicles," *IEEE Transactions on Smart Grid*, vol. 4, no. 1 (2013): 596–605.

[42] Mal, Siddhartha, Arunabh Chattopadhyay, Albert Yang, and Rajit Gadh. "Electric vehicle smart charging and vehicle-to-grid operation," *International Journal of Parallel, Emergent and Distributed Systems*, vol. 28, no. 3 (2013): 249–265.

[43] Deilami, Sara, Amir S. Masoum, Paul S. Moses, and Mohammad AS Masoum. "Real-time coordination of plug-in electric vehicle charging in smart grids to minimize power losses and improve voltage profile," *IEEE Transactions on Smart Grid*, vol. 2, no. 3 (2011): 456–467.

[44] Lam, Ka Lun, King Tim Ko, Hoi Yan Tung, Hoi Ching Tung, Wah Ching Lee, Kim Fung Tsang, and Loi Lei Lai. "Advanced metering infrastructure for electric vehicle charging," *Smart Grid and Renewable Energy*, vol. 2, no. 4 (2011): 312–323.

[45] Lim, Yujin, Hak-Man Kim, and Sanggil Kang. "Information system for electric vehicle in wireless sensor networks," In *International Conference on Future Generation Communication and Networking*, Berlin, Heidelberg: Springer, 2010, pp. 199–206.

[46] Yilmaz, Murat, and Philip T. Krein. "Review of benefits and challenges of vehicle-to-grid technology," In *2012 IEEE Energy Conversion Congress and Exposition (ECCE)*, IEEE, 2012, pp. 3082–3089.

[47] Habib, Salman, Muhammad Kamran, and Umar Rashid. "Impact analysis of vehicle-to-grid technology and charging strategies of electric vehicles on distribution networks–a review," *Journal of Power Sources*, vol. 277 (2015): 205–214.

[48] Online: https://fame2.heavyindustry.gov.in/

[49] Online: https://transport.delhi.gov.in/sites/default/files/AllPDF/Delhi_Electric_Vehicles_Policy_2020.pdf

[50] Online: http://udyogbandhu.com/DataFiles/CMS/file/Electrical%20%20vehicle%20policy_english_Aug7_2

[51] Verzijlbergh, Remco A., Laurens J. De Vries, and Zofia Lukszo. "Renewable energy sources and responsive demand. Do we need congestion management in the distribution grid?," *IEEE Transactions on Power Systems*, vol. 29, no. 5 (2014): 2119–2128.

[52] Liu, Liansheng, Fanxin Kong, Xue Liu, Yu Peng, and Qinglong Wang. "A review on electric vehicles interacting with renewable energy in smart grid," *Renewable and Sustainable Energy Reviews*, vol. 51 (2015): 648–661.

[53] Un-Noor, Fuad, Sanjeevikumar Padmanaban, Lucian Mihet-Popa, Mohammad Nurunnabi Mollah, and Eklas Hossain. "A comprehensive study of key electric vehicle (EV) components, technologies, challenges, impacts, and future direction of development," *Energies*, vol. 10, no. 8 (2017): 1217.

[54] Fortes, Sergio, José Antonio Santoyo-Ramón, David Palacios, Eduardo Baena, Rocío Mora-García, Miguel Medina, Patricia Mora, and Raquel Barco. "The campus as a smart city: University of Málaga environmental, learning, and research approaches," *Sensors*, vol. 19, no. 6 (2019): 1349.

[55] R. Redl, "Power electronics and electromagnetic compatibility," In *PESC Record. 27th Annual IEEE Power Electronics Specialists Conference*, vol. 1 (1996): 15–21, doi:10.1109/PESC.1996.548553

Chapter 3

Implementation issues with large-scale renewable energy sources and electric vehicle charging stations on the smart grid

M. Jayachandran
Sri Manakula Vinayagar Engineering College, Puducherry, India

C. Kalaiarasy and C. Kalaivani
Puducherry Technological University, Puducherry, India

CONTENTS

3.1 INTRODUCTION

High renewable energy source (RES) penetration into established power networks can cause reliability and stability concerns owing to its intermittency, causing uncertainty in the operation of the power system. In this scenario, storage solutions for renewable power generation provide auxiliary services to alleviate the supply–demand imbalance. Few studies have looked

DOI: 10.1201/9781003311195-3

at the control of power electronic converters for grid-interface solar power generation in-depth [1–2]. In addition, sophisticated control techniques and technologies for Voltage/frequency regulation in sustainable microgrids are being investigated in the literature [3].

From a clean energy transportation perspective, increased power generation from carbon-free RES would be enabled through the electrification of the transportation infrastructure. The integration of EV and RES technologies will provide a practical solution to the usage of renewables for future transportation while also ensuring power grid stability. The coordinated control approach of electric mobility and renewables has been proposed in the literature in order to achieve better frequency regulation. Many emerging technologies have been used in the contemporary era to connect massive EVs with electrical networks. Vehicle to grid (V2G) technology enables communication between the distribution grid and EVs so as to assure the power grid's stability and sustainability. Furthermore, the V2G approach offers dedicated support to the power grid, such as voltage and frequency correction, as well as a spinning reserve. However, the design of an EV charging infrastructure for a stable electricity supply is a critical factor in promoting EV demand. According to recent research, the EV charger's flexible operation can interact with power distribution grids, microgrids, and smart homes [4].

The integration of renewables into the electricity system has raised both planning and operational challenges. The smart grid's effective integration of distributed generation (DG) improves generating capacity, electricity quality, and dependability. Recent studies have also considered other aspects, including renewable penetration into the smart grid, resource planning, and demand-side management (DSM) [5].

3.2 LARGE-SCALE RENEWABLE INTEGRATION INTO THE SMART GRID

Heavy electrical loads, such as EVs, might cause energy demands during peak times, having an impact on grid stability. To address this issue, high-performance renewable energy resources are being incorporated into the electric grid to build a more sustainable, self-sufficient, and clean energy system. The addition of renewable resources to the smart grid enables surplus electricity to be injected into the power system and tends to minimize carbon pollutions. Figure 3.1 summarizes the key concerns with connecting large-scale renewable resources to the electricity distribution grid [6]. However, grid operators face significant hurdles because of the unpredictability and fluctuation of RES. To counter these concerns, large-scale battery storage has been installed to mitigate the variability between supply and demand. According to recent findings, high photovoltaic penetration into utility grids is intended to operate without any degradation of reliability and power quality [7]. They also offer grid services, including active/reactive

Renewable energy integration issues	Solution Methodologies
The intermittent power generation of renewable source dependent on weather, season, and time of day	Accurate generation forecasting and power balancing between energy resources are needed
Many of renewable sources do not have reactive power generation	Voltage and frequency control are necessary
The inertia-less generation such as solar or sudden generation loss may lead to system instability	Good inertia and primary frequency control are required to hold and improve the frequency during sudden generation loss contingency
Power quality issues are also within the scope including harmonics, flicker, and under voltage ride through capability	It should be maintained according to IEEE and IET standard to normalize the system operation
Power management and maximum power point tracking (MPPT)	Requires proper converters and controls

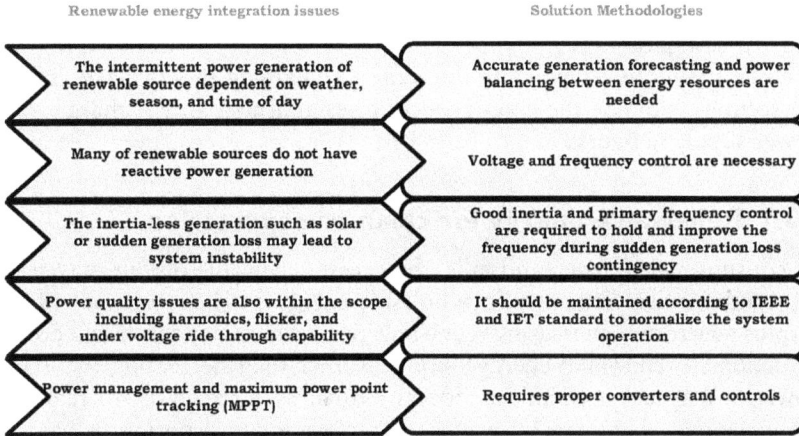

Figure 3.1 Concerns about renewable energy integration in smart grids and their solutions.

power regulation. Particularly, smart inverters in decentralized renewable plants, as synchronous machines, provide frequency regulation using solar-storage technologies [8].

The term "Distributed Energy Resources (DERs)" refers to a category of energy resources that includes PV generators, synchronous generators, wind energy conversion systems, fuel cells, and battery technologies. They can generate, convert, and store energy, as well as reconnecting to the power grid. Dispatchable and controllable DERs with a communication infrastructure are required for large-scale DER deployment with minimum grid effect. Establishing such devices for a sustainable electricity network necessitates achieving the following features: (i) renewable generator scheduling, supervision, and management; (ii) instantaneous power control; (iii) voltage regulation to preserve synchronization; and (iv) ancillary services, including load regulation, spinning and non-spinning reserve, energy loss reduction, and so on [9]. Future research directions include the interconnection of renewables with data centres, internet services, transportation, and cellular networks. Renewable resources would also provide cognitive radio between smart grid devices, and communication needs. Other essential aspects of a future renewable-supported smart grid include information security and user privacy.

3.3 THE IMPACT OF EVCS ON THE SMART DISTRIBUTION NETWORK

AC level 2 chargers are an essential component of EV charging points. EV charging stations, several types of the energy storage systems (ESS), such as flywheels, batteries, and hybrid energy storage devices, can be used.

The effect of these storage devices on electric vehicle chargers is currently under investigation [10]. A supercapacitor is additionally installed in plug-in electric vehicles to expedite the battery's charging process. The following sections examine the consequences of establishing an EV charger on a power supply network.

3.3.1 Renewable energy for clean transportation

As consumer power demand falls, the excess renewable output may result in negative price frequency in wholesale energy markets. Because of this surplus generation, increasing renewable penetration may result in a greater reduction in renewable energy output. Rather than decreasing the excess power, it may use a chemical process known as 'electrolysis' to retain all the generated electricity by way of hydrogen for transportation benefits. Negative pricing from green sources on the grid promotes more eco-friendly energy requirements, like fuel cell vehicles (FCV). As a result, progression in hydrogen storage for FCVs in the automobile sector must continue to focus on FCV range, flexibility, and fast charging [11].

3.3.2 EVCS planning for electric vehicles

With suitable energy policies and rates, EV aggregators can support supplementary services in the wholesale energy marketplace. Long recharging times and the restricted driving range for EVs have posed serious barriers to the expansion of the EV market. It has been suggested that the installation of a large onboard battery with a fixed charging capacity might alleviate these problems. Figure 3.2 shows DC/AC conductive as well as capacitive/inductive wireless power transfer (CPT/IPT) chargers are two types of stationary chargers that may be categorized according to the conduction medium. On the other side, large onboard batteries raise the vehicle's price, scale, and energy consumption. It is also necessary to have a fast charger [12]. Furthermore, quasi-dynamic/dynamic charging mechanisms are often

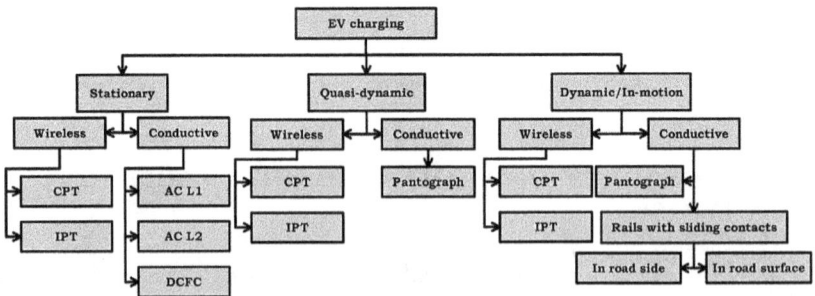

Figure 3.2 Chargers for electric vehicles.

exploited for in-motion fast charging to enable low-cost charge-sustaining operation, allowing EVs to charge while moving and transiting. These technologies are being investigated in order to increase EVs' achievable driving distances while simultaneously permitting the utilization of tiny onboard battery banks. Interstate freeways with limited-speed driving corridors, as well as crossroad bus stops, may benefit from quasi-dynamic charging. Dynamic charging, on the other hand, appears to be the best fit for high-speed freeways. Both systems are capable of operating in conductive (or) wireless modes. The charge-sustaining operation may be realized with the proper development of wireless dynamic charging infrastructure for hybrid vehicles on roadways [13].

3.3.3 EV fast charging station with multiple energy sources

Interoperability is described as the capability of two or more devices to participate in a variety of distributed applications. Despite the interoperability of the dynamic charging system, the system's road span and system power are completely contradictory. Large-power systems will minimize road coverage, but they need powerful modules that are expensive. Recently, research articles on alternative EV system topologies in utility grids have been published [14]. However, the research has yet to be thoroughly examined. Charging batteries during off-peak hours is desirable in the case of grid-connected operation. Grid-connected plugin vehicles that can be quickly recharged with additional power from photovoltaics and storage devices have become common in recent years. However, this technique necessitates the procurement of high-power rating modules, which are costly. In recent decades, the modular multi-port converter design has been introduced to facilitate simultaneous and continuous charging of vehicles at EVCS utilizing low-power circuits to investigate this issue [15–16]. For future work, an enhanced control mechanism must be established.

3.3.4 Managing EV charging an EVCS

To alleviate range anxiety in electric vehicles, one such primary difficulty in raising the contribution of EVs to the utility grid is the integration of an EVCS into existing transportation and electricity networks. Recent analysis reveals that the best location and capacity of the EVCS are decided by an optimization algorithm that takes into account the area's traffic patterns [17]. The demand for electric mobility and the size of the population can each have a significant impact on power and transportation networks. The EV driver, the grid, and the service provider are all involved in EV charging management within the EVCS. Innumerable charging stations continue to offer service providers with access to manage plug-in hybrid vehicle recharges, reduce grid load, and increase profit [18].

In crowded cities, an abundance of vehicles is demanding that their batteries be recharged concurrently at one particular outlet. To mitigate this obstacle, researchers have demonstrated a battery-swapping station for quickly replacing a drained battery with a recharged battery pack. When compared to EVCS, this approach takes a few minutes and provides greater comfortability and adaptability [19]. Another possible issue in the V2G application is the incorporation of an energy management system for grid-connected hybrid vehicles [20]. Furthermore, security and protection concerns in the huge V2G infrastructure, as well as their implications in terms of grid stability, need to be addressed in future studies.

3.4 RENEWABLES IN SMART GRIDS: PLANNING AND OPERATION

The goal of capacity planning in the distributed generator is to determine whether the present system should be upgraded or additional resources should be acquired to satisfy electricity needs. In order to get the optimum solution for reliable operation, the role of optimization in the distributed generation has gained considerable attention. Interventions in distribution grid planning and operation with high penetration of renewables and storage technologies can optimize the power flow among grid, storage, and dispersed loads, ensuring electric supply reliability and cost. To date, only a few renewable–nuclear combination research investigations have been found. A nuclear–renewable hybrid energy system (N-RHES) generates enough electricity to fulfill load needs. However, capital costs, construction time, environmental safety, the lifespan of the project, scalability, site acquirement, and technology hurdles are among the issues to be taken into consideration when constructing economical and high-intensity systems. The mix of small-scale nuclear–renewable power facilities is suitable for distant customers. They offer numerous choices for connecting the system's input and output. When demand exceeds generation, these systems import power from the grid. Conversely, when demand is low, excess energy may be preserved in a storage system or supplied to the grid.

This system enables realistic energy-saving options for workplaces, large corporations, and institutions. The coupling approach in a renewable–nuclear combination can be characterized as either tightly or loosely coupled resources based on the availability of renewable resources and load demand. As shown in Figure 3.3, the energy generated by grid-connected tightly coupled nuclear–renewable is used to meet electric demand, with any excess energy being sent back into the grid. This technology can supply electricity to large industrial loads like hydrogen loads. To maximize system efficiency, subsequent research on renewable–nuclear should elaborate on the dynamic control of large-scale storage systems. The effective integration

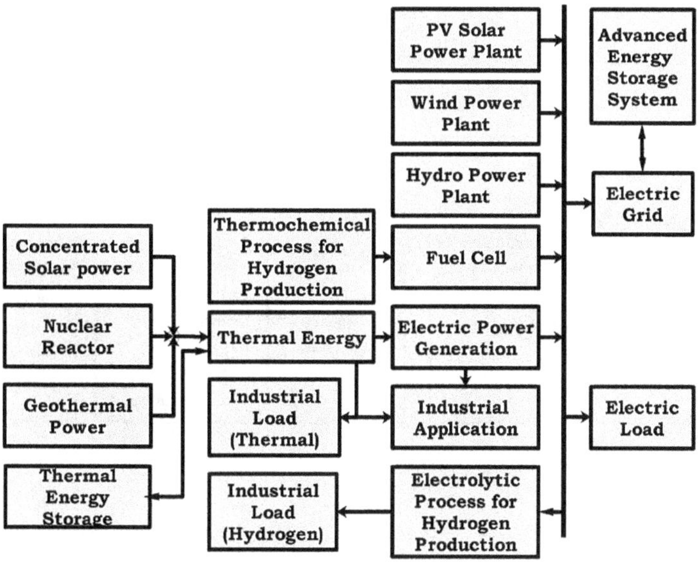

Figure 3.3 Grid-connected nuclear-renewable micro-hybrid energy system.

of different renewable resources into a smart grid can address the environmental and economic power dispatch challenge [21].

3.4.1 Energy balance and power continuity

Renewable distributed generation and storage play an essential element in ensuring power balance and continuity in an interconnected network. The influence of the smart grid on increasing power quality can deliver clean electricity to end-users with allowable disturbance. Voltage and frequency fluctuations, voltage transients, waveform distortion, harmonic current, and surge current all affect utilities and network users. Figure 3.4 illustrates a variety of terms related to electric power quality on both the consumer and supplier sides [22].

3.4.2 Improved power quality with smart inverters

In recent decades, a substantial quantity of literature on power quality control has been published [23]. These investigations have mostly concentrated on compensation on unbalanced and harmonic voltages, as seen in Figure 3.5. The power balance is disrupted in the distribution system due to the rising penetration of renewables, causing reverse power flows during brownouts. In this context, islanded mode predictions are utilized with machine learning algorithms to alert PV inverter control. With the help of

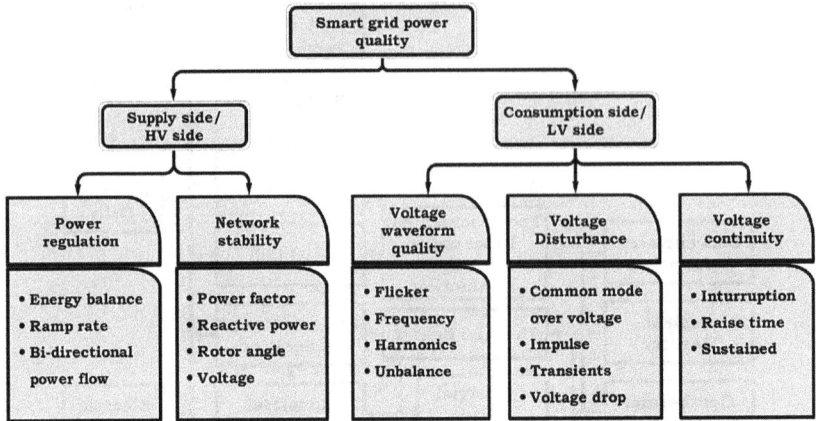

Figure 3.4 Power quality issues in smart grid.

the command signal, the control circuitry isolates the inverter and shifts to a V/f regulation mode to offer secure reliable electricity to the islanded region. Smart inverters in distributed generation can regulate the power generation according to network frequency and point of common coupling voltage, according to the IEEE 1547–2018 standard. Responding to grid circumstances, smart inverters may offer dynamic reactive power over a range of power factor values. To enhance power quality in the renewable generation system, smart inverters communicate with utilities through the energy management system and aggregator. In addition, smart inverters give operational status to assist the utility's distribution system.

3.4.3 Power quality management in smart grid

To manage and monitor power quality issues, smart grid technologies, such as an advanced metering infrastructure, distribution automation, geographic information systems, volt/var control systems, power quality analyzers, and supervisory control and data acquisition, are installed into the electricity network. The energy trading interface uses advanced metering infrastructure to interact with other domains [24]. Further research on power quality enhancement in smart grids should focus on the widespread implementation of sensing, processing, and communication networks [25].

3.4.4 SMART principles in the distribution system

As shown in Figure 3.6, smart principles are used in grids with high renewable penetration. Their role is to monitor, compute the detected value, and convey a decision as quickly as possible. The smart concepts of a planning tool, brown box, inertia controller, storage manager, and signalling

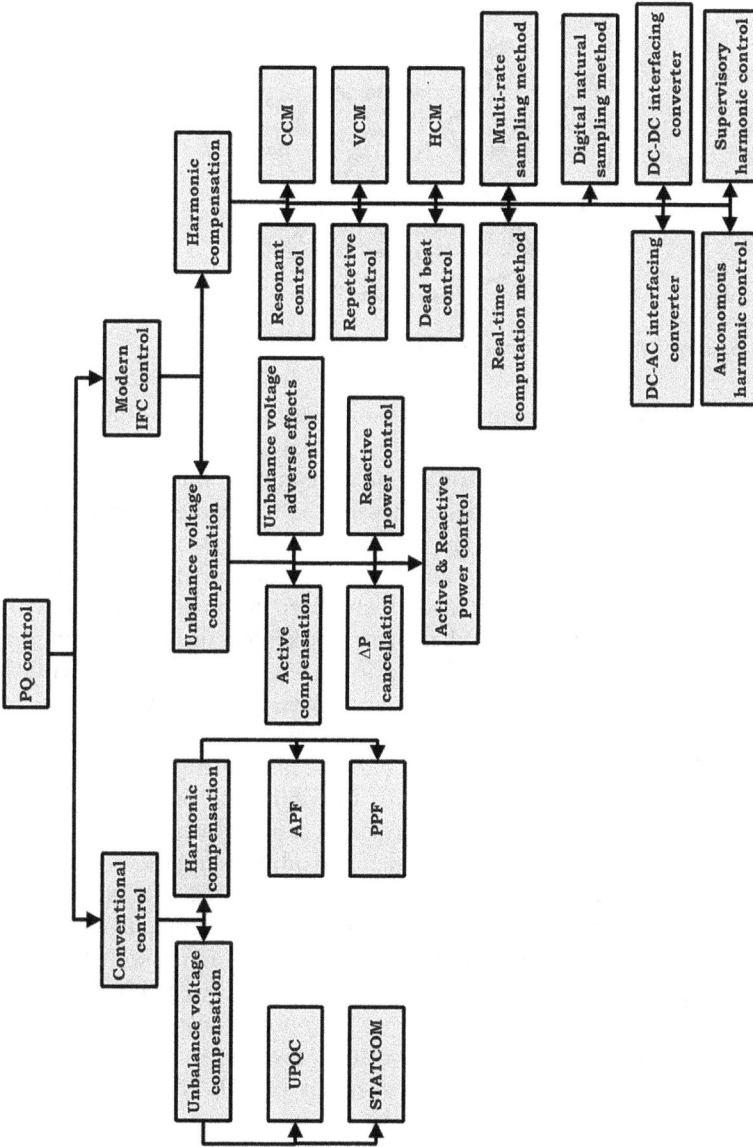

Figure 3.5 Power quality control of the smart grid.

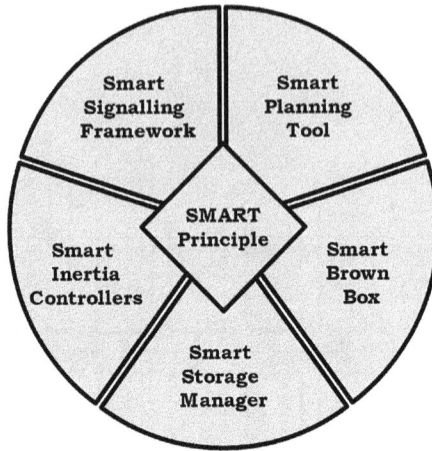

Figure 3.6 Smart principles for planning and operation of grids.

framework can assist distribution system operators in upgrading their traditional grids. The smart inertia controller emulates inertia from renewables to the grid for frequency regulation. To provide inertia equivalent to 5–10% of the capacity, the inertia on-demand with the hardware-in-loop concept should be examined in future. The smart brown box is designed to transform blackouts into brownouts amid a supply deficit. In such instances, utilities often use rolling blackouts, which alternate power outages across various feeds. Customers that utilize brown boxes, on the other hand, may have their usage lowered through load prioritization in conjunction with the utility, causing a reduced supply in service feeders. A peak shaving approach is devised for these systems to reduce peak and reserve capacity investments. When it comes to energy sharing, the battery may be employed as a common source for utilities that are correctly distributed across several induces to facilitate flexible control. Such approaches can uncover the actual value of communal storage [26]. To achieve demand adjustment, the smart brown box deployed at each end-location user prioritizes customer loads. They are supposed to enable adaptation based on consumer type. Under a generation deficiency situation, the problems and techniques for leveraging and fostering load prioritizing should be investigated to freely curtail the load. Algorithms must be established for the optimal allocation of limited resources at the energy supplier and consumer levels.

Furthermore, smart storage mechanisms may be utilized to govern utility storage systems to supplement renewable energy without placing an additional burden on the network. As a consequence, efficient intervention utilization requires sufficient planning and appropriate real-time control. A smart planning tool may also assist in deciding renewable and storage investment, creating energy exchange communities, guiding policies, and designing tariff systems. The SMART storage manager may manage storage on either the supply side or the demand side. The operational policy is produced as a supply-side supplement relying on renewable resource location, grid dispatch command, and storage devices. Energy-sharing regulations and other trade flows regulate storage functions since they are demand-side resources. More research is needed to establish the efficacy of dynamic control in storage systems in dictating charging/discharging policies to correct renewables' unpredictability and ensure energy access to all customers.

Finally, a smart signalling architecture connects the smart brown box, inertia controller, and storage manager in order to send commands to the generator, estimating the state of charge of the battery, and calculating end-use consumption. They also serve as the cornerstone for operational control, bringing together all of the options for easing the demand for renewables. The design of a smart signalling framework should use optimized control techniques and stochastic systems to coordinate generator, load, and storage resources in normal and emergency scenarios [27–28].

3.5 FUTURE AREAS OF RESEARCH IN LARGE-SCALE RES AND EV IMPLEMENTATIONS

This chapter highlights recent developments and significant challenges in the smart grid integration of RES and EVCSs on a large scale [28–29]. The following are some potential future advancements in the modernization of power grids.

3.5.1 RES integration

Forecasting, scheduling, and integrating renewables with Supervisory Control and Data Acquisition (SCADA) should be the focus of future smart grid research. Another promising area of study is the installation of Advanced Metering Infrastructure (AMI) with price incentives for demand-side management, the expansion of balancing areas, the testing of alternative battery technologies, the enhancement of situational awareness, improved visualization, and the steady functioning of the grid.

3.5.2 Intelligent charging stations

Future studies should concentrate on developing a multiport DC fast charger with a 510kW for electric mobility applications. In addition, every autonomous vehicle will require 6.6kW AC and 36kW DC fast chargers in the future. The EV charging infrastructure needs to be as cost-effective as possible while still being reliable and controllable.

As a result, the optimum battery charging technique for multi-car charging is of the areas that will be the subject of future study. Enabling communication and security capabilities into electric vehicle chargers to support intelligent and safe EV charging systems might be another topic of future study. Furthermore, flash charging technology, which reduces recharge time at auxiliary feeding points, is a growing concept in next-generation automobiles. This takes advantage of the onboard converter to give a simple and robust rapid charging interface. Through incorporated peak shaving functionality, this platform allows an eco-friendly public transportation system with minimal connectivity and electricity prices.

3.5.3 Smart operational planning in a power network

More study is required to build a game-theoretic smart planning tool and a data-driven analysis for capacity planning, sizing, addition, and expansion of the network. Furthermore, mixed-energy systems are the most cost-effective solution in terms of increasing grid flexibility. In the future, several instances and case studies of mixed-energy resources should be investigated, with an emphasis being placed on the interaction between planning, operations, and cost of the smart grid [29].

REFERENCES

[1] L. Ashok Kumar, S. Albert Alexander, and Madhuvanthani Rajendran, "Chapter 1 – Inverter topologies for solar PV," *Power Electronic Converters for Solar Photovoltaic Systems*, Academic Press, Cambridge, MA, pp. 1–39, 2021.

[2] M. Shahbazi, and A. Khorsandi, "Chapter 10 – Power electronic converters in microgrid applications," *Microgrid*, Butterworth-Heinemann, Oxford, pp. 281–309, 2017.

[3] Y. Han, H. Li, P. Shen, E. A. A. Coelho, and J. M. Guerrero, "Review of active and reactive power sharing strategies in hierarchical controlled microgrids," *IEEE Transactions on Power Electronics*, vol. 32, no. 3, pp. 2427–2451, 2017.

[4] S. Amamra, and J. Marco, "Vehicle-to-grid aggregator to support power grid and reduce electric vehicle charging cost," *IEEE Access*, vol. 7, pp. 178528–178538, 2019.

[5] L. Bhamidi, and S. Sivasubramani, "Optimal planning and operational strategy of a residential microgrid with demand side management," *IEEE Systems Journal*, vol. 14, no. 2, pp. 2624–2632, 2020.

[6] L. H. Koh, Y. K. Tan, P. Wang, and K. J. Tseng, "Renewable energy integration into smart grids: Problems and solutions – Singapore experience," In *2012 IEEE Power and Energy Society General Meeting*, 2012, pp. 1–7.

[7] K. N. Nwaigwe, P. Mutabilwa, and E. Dintwa, "An overview of solar power (PV systems) integration into electricity grids," *Materials Science for Energy Technologies*, vol. 2, no. 3, pp. 629–633, 2019.

[8] Abdul Motin Howlader, Staci Sadoyama, Leon R. Roose, and Yan Chen, "Active power control to mitigate voltage and frequency deviations for the smart grid using smart PV inverters," *Applied Energy*, vol. 258, p. 114000, 2020.

[9] Sherif M. Ismael, Shady H.E. Abdel Aleem, Almoataz Y. Abdelaziz, and Ahmed F. Zobaa, "State-of-the-art of hosting capacity in modern power systems with distributed generation," *Renewable Energy*, vol. 130, pp. 1002–1020, 2019.

[10] Snigdha Sharma, Amrish K. Panwar, and M.M. Tripathi, "Storage technologies for electric vehicles," *Journal of Traffic and Transportation Engineering*, vol. 7, no. 3, pp. 340–361, 2020.

[11] M.D. Paster, R.K. Ahluwalia, G. Berry, A. Elgowainy, S. Lasher, K. McKenney, and M. Gardiner, "Hydrogen storage technology options for fuel cell vehicles: Well-to-wheel costs, energy efficiencies, and greenhouse gas emissions," *International Journal of Hydrogen Energy*, vol. 36, no. 22, pp. 14534–14551, 2011.

[12] Chirag Panchal, Sascha Stegen, and Junwei Lu, "Review of static and dynamic wireless electric vehicle charging system," *Engineering Science and Technology, an International Journal*, vol. 21, no. 5, pp. 922–937, 2018.

[13] Yaseen Alwesabi, Yong Wang, Raul Avalos, and Zhaocai Liu, "Electric bus scheduling under single depot dynamic wireless charging infrastructure planning," *Energy*, vol. 213, p. 118855, 2020.

[14] Dai-Duong Tran, Majid Vafaeipour, Mohamed El Baghdadi, Ricardo Barrero, Joeri Van Mierlo, and Omar Hegazy, "Thorough state-of-the-art analysis of electric and hybrid vehicle powertrains: Topologies and integrated energy management strategies," *Renewable and Sustainable Energy Reviews*, vol. 119, p. 109596, 2020.

[15] R. Palanisamy, J. Seenithangam, and R. Palanisamy, "A hybrid output multiport converter for standalone loads and photovoltaic array integration," *International Transactions on Electrical Energy Systems*, vol. 30, p. 12410, 2020.

[16] P. Rajan, and S. Jeevananthan, "A new partially isolated hybrid output multiport multilevel converter for photovoltaic based power supplies," *Journal of Energy Storage*, vol. 45, p. 103436, 2021.

[17] Liang Chen, Chunxiang Xu, Heqing Song, and Kittisak Jermsittiparsert, "Optimal sizing and sitting of EVCS in the distribution system using metaheuristics: A case study," *Energy Reports*, vol. 7, pp. 208–217, 2021.

[18] Y. Susowake, H. Yongyi, T. Senjyu, A. M. Howlader, and P. Mandal, "Optimum operation plan for multiple existing EV charging stations," In *2018 IEEE PES Asia-Pacific Power and Energy Engineering Conference (APPEEC)*, 2018, pp. 611–615.

[19] A. Relan, V. Gupta, R. Kumar, S. Vyas, and B. K. Panigrahi, "Optimal siting of electric vehicle battery swapping stations with centralized charging," In *2020 IEEE International Conference on Power Electronics, Drives and Energy Systems (PEDES)*, 2020, pp. 1–6.

[20] Wajahat Khan, Furkan Ahmad, and Mohammad Saad Alam, "Fast EV charging station integration with grid ensuring optimal and quality power exchange," *Engineering Science and Technology, an International Journal*, vol. 22, no. 1, pp. 143–152, 2019.

[21] Muhammad R. Abdussami, Md Ibrahim Adham, and Hossam A. Gabbar, "Modeling and performance analysis of nuclear-renewable micro-hybrid energy system based on different coupling methods," *Energy Reports*, vol. 6, no. 6, pp. 189–206, 2020.

[22] Amalorpavaraj Rini Ann Jerin, Natarajan Prabaharan, Nallapaneni Manoj Kumar, Kaliannan Palanisamy, Subramaniam Umashankar, and Pierluigi Siano, "10 – Smart grid and power quality issues," *Woodhead Publishing Series in Energy, Hybrid-Renewable Energy Systems in Microgrids*, Woodhead Publishing, Cambridge, pp. 195–202, 2018.

[23] Tao Jin, Yueling Chen, Jintao Guo, Mengqi Wang, and Mohamed A. Mohamed, "An effective compensation control strategy for power quality enhancement of unified power quality conditioner," *Energy Reports*, vol. 6, pp. 2167–2179, 2020.

[24] M. Jayachandran, C. Reddy, S. Padmanaban, and A. Milyani, "Operational planning steps in smart electric power delivery system," *Scientific Reports*, vol. 11, no. 1, pp. 1–21, 2021.

[25] M. Jayachandran, and C. Kalaiarasy, "Power-domain NOMA for massive connectivity in smart grid communication networks," *Proceedings of International Conference on Power Electronics and Renewable Energy Systems*. Springer, pp. 205–212, 2022.

[26] W. Tushar, B. Chai, C. Yuen, S. Huang, D. B. Smith, H. V. Poor, and Z. Yang, "Energy storage sharing in smart grid: A modified auction-based approach," *IEEE Transactions on Smart Grid*, vol. 7, no. 3, pp. 1462–1475, 2016.

[27] D. R. Basina, S. Kumar, S. Padhi, A. Sarkar, A. Mondal, and R. Krithi, "Brownout based blackout avoidance strategies in smart grids," *IEEE Transactions on Sustainable Computing*, vol. 6, pp. 586–598, 2020.

[28] A. Ramanujam, M. Parihar, S. Swain, and K. Ramamritham, "Design and development of brownout control strategy using end-point load control," In *Proceedings of the Eleventh ACM International Conference on Future Energy Systems*, 2020, pp. 293–298.

[29] M. Jayachandran, K. Prasada Rao, Ranjith Kumar Gatla, C. Kalaivani, C. Kalaiarasy, C. Logasabarirajan, "Operational concerns and solutions in smart electricity distribution systems," *Utilities Policy*, vol. 74, p. 101329, 2022.

Chapter 4

Analysis of a fuel cell-fed BLDC motor drive with a double boost converter for electric vehicle application

K. Kumar and V. Lakshmi Devi
Sri Venkateswara College of Engineering (SVCE), Tirupati, India

Avagaddi Prasad
Sasi Institute of Technology & Engineering, Tadepalligudem, India

Hanumantha Reddy Gali
Sri Venkateswara Engineering College (SVEC), A.P., India

Ramji Tiwari
Sri Krishna College of Engineering and Technology, Kuniamuthur, India

CONTENTS

4.1 INTRODUCTION

In recent decades, there have been some growing concerns regarding the emission of greenhouse gases and the depletion of fossil fuel reserves which have made fuel cell energy sources increasingly attractive. This is mainly due to their high reliability, low levels of pollutant emission and minimal maintenance. Over the past decade, the transport emissions have rapidly

DOI: 10.1201/9781003311195-4

59

increased at a faster rate than has occurred in any other energy sector. This is because of the transportation section is seen as responsible for 36%–78% of the main components of urban air pollution, 28% of all US greenhouse gas emissions, 34% of all carbon dioxide emissions, and 68% of all oil consumption. While the level of world transportation is increasing in line with economic growth, gas emissions will also increase. Fossil fuel demand also raises up many concerns and challenges, such as climate change, CO_2 emissions, and supply cost increases [1].

Fuel cells are electrochemical devices that convert chemical energy directly into electrical energy through an electrolytic reaction, emitting only heat and water. In spite of being a clean source of energy, they are only capable of producing unregulated DC voltage, hence the need for power converters to interface the driven load. Fuel hydrogen is the preferred choice for the manufacturers of EVs [2].

Fuel cells and batteries are the two primary choices to power electric vehicles, boats, and even ships. Both of these sources are used to generate electricity that drives electric motors and eliminates pollution. Fuel cells generate their energy from hydrogen that is stored in the cylinders, whereas batteries can be charged or recharged from the electrical grid. Hydrogen and electricity from the grid can be made with a very low level of carbon emission, including renewable energy sources. Batteries are not always the long-term solution for energy storage as they provide short-term solutions and dangerous pollutants and significant waste concerns. Hydrogen appears to be the only valid clean source for energy storage [3].

The recent emergence of the hydrogen (H_2) fuel cell electric vehicle (FCEV) guarantees the benign nature of the transportation industry. Several companies, including Toyota and Hyundai, have begun to commercialize H_2 FCEV with relatively comparable properties to meet the demands of the renewable energy-based future. For instance, the Hyundai Tucson can drive around 400 miles after just 3 minutes of a simple H_2 charge. Additionally, to keep abreast of supporting H_2 FCEVs, Germany has decided to help construct the fueling stations by supporting funding and restricting the production of oil-based vehicles [4].

The main advantage of the hybrid electric vehicle, in addition to its improved fuel economy, is the reduction in carbon dioxide emissions. In May 2016, the carbon dioxide emission rate was near 407.70 ppm, and it is increasing gradually [5]. Hybrid electric vehicles use many state-of-the-art environment-friendly technologies, such as a regenerative braking system, fuel cell etc., which will greatly reduce the emission of those harmful elements which are responsible for environment pollution [6]. Along with these green technologies, the use of efficient power electronics and lighting systems will greatly improve the overall performance of hybrid electric vehicles.

In recent times, there has also been a rapid growth in the use of fuel cell technology and it shows its great potential as a green energy source in the near future because of its unique features (low emission of pollutants, fuel flexibility, and modular structure). However, these renewable energy sources are becoming particularly popular even though they are limited by the capital costs involved and the intermittency in power production.

Fuel cells and electric motors operate on different voltages, and a DC/DC converter is needed to allow for this discrepancy. Many conventional design topologies include multiple DC/DC converters that increase total mass, cost and complexity. A new design topology to allocate power is developed to give higher efficiency with a difference of consolidating two or three DC/DC converters into one. The battery and fuel cell operate in certain regimes that lower losses, and the depth of a battery's discharge operates for a longer life of the battery. Therefore, we offer a model, design and control of a fuel cell-powered scooter in urban use that can reduce a battery pack's maintenance or replacement cost through ensuring a longer life span and greater range.

Global warming and environmental pollution are no longer scientific warnings; they are a part of everyday life in the 21st century. The weather in many countries is characterized by record-breaking temperatures and precipitation, with droughts occurring almost every season. In addition, highly industrialized cities continue to suffer from deteriorating air quality and the related health issues. Although the debate remains about these environmental anomalies being solely the result of human activities, no one can deny the effect of highly concentrated chemicals that are emitted from vehicles and manufacturing plants, which are rarely observed in nature (with the possible exception of those near volcanoes). Researchers and engineers continue to attempt to explain the effects of human activity on the earth, and, more importantly, to find solutions to this problem which will impact on both current and future generations. However, the moral idea that the earth should be passed on as received is frequently superseded by the selfish notion of making instant profits from burning fossil fuels. Hence, the transition from fossil fuels to renewable energy sources is slow and sometimes requires government enforcement to accelerate. One notable landmark of the United Nations (UN) in this regard is the 2015 Paris Agreement. Signed by 196 countries, including those most responsible for greenhouse gas (GHG) emissions, the agreement is a step in the right direction in terms of reducing climate change. According to the agreement, the parties must reduce GHG emissions to limit global warming to under 2°C compared with the pre-industrial level.

In recent years, the proportion of energy consumption from renewable energy resources is increasing compared to that of conventional sources due to the surge in demand for clean power. Environmental conferences predict that fossil fuels will be totally depleted by the year 2050. So we have to reserve these sources for next generations and give an increased importance to the use of renewable resources in generating electrical energy [7]. Renewable energy resources include photovoltaic systems, wind energy systems, fuel cells and others. Fuel cells are efficient and reliable and they can easily generate on-board electricity for automobiles [8].

In this chapter, an EV system is designed with FC as an energy source, a double boost DC–DC converter accompanying a PSO-based MPPT controller and a BLDC motor drive controlled by the Hall Effect signal method. The proposed system is developed in a MATLAB/Simulink model with variable FC input temperature conditions and a comparative analysis is done with different MPPT controllers. Section 4.2 describes the proposed system for EV application; Section 4.3 describes the control techniques involved in DC–DC converter and in Inverter part; and Section 4.4 presents results and the discussion section followed by conclusions made in the present research study.

4.2 PROPOSED FUEL CELL FED ELECTRIC VEHICLE SYSTEM DESIGN

Figure 4.1 depicts the proposed circuit of the fuel cell-fed electric vehicle system with a double boost converter. This system achieves double output voltage by using a conventional boost and a switched capacitor. The double boost converter is used to provide a double output voltage without operating it in extreme conditions.

4.2.1 Fuel cell

A fuel cell is a static device, which converts stored chemical energy into electrical energy. By arranging the anode, electrolyte and cathode with the help of anode and cathode catalyst in either series or parallel, a fuel cell is constructed to produce electrical energy through a chemical reaction between the hydrogen and oxygen [9–11]. The fuel cell characteristics of the proposed system are illustrated in Figure 4.2.

In the anode, the hydrogen (H) undergone into molecules as given in Equation (4.1)

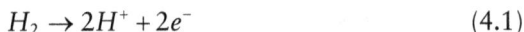

$$H_2 \rightarrow 2H^+ + 2e^- \tag{4.1}$$

In the cathode, the chemical reaction is as given in Equation (4.2)

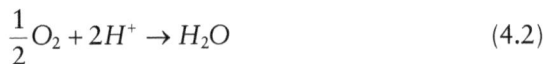

$$\frac{1}{2}O_2 + 2H^+ \rightarrow H_2O \tag{4.2}$$

The voltage generated from the single fuel cell is derived from the Nernst Equation [10], as given in Equation (4.3)

$$E_{cell} = E_O + \frac{RT}{2F}\ln\frac{PH_2\sqrt{PO_2}}{PH_2O} \tag{4.3}$$

Where,

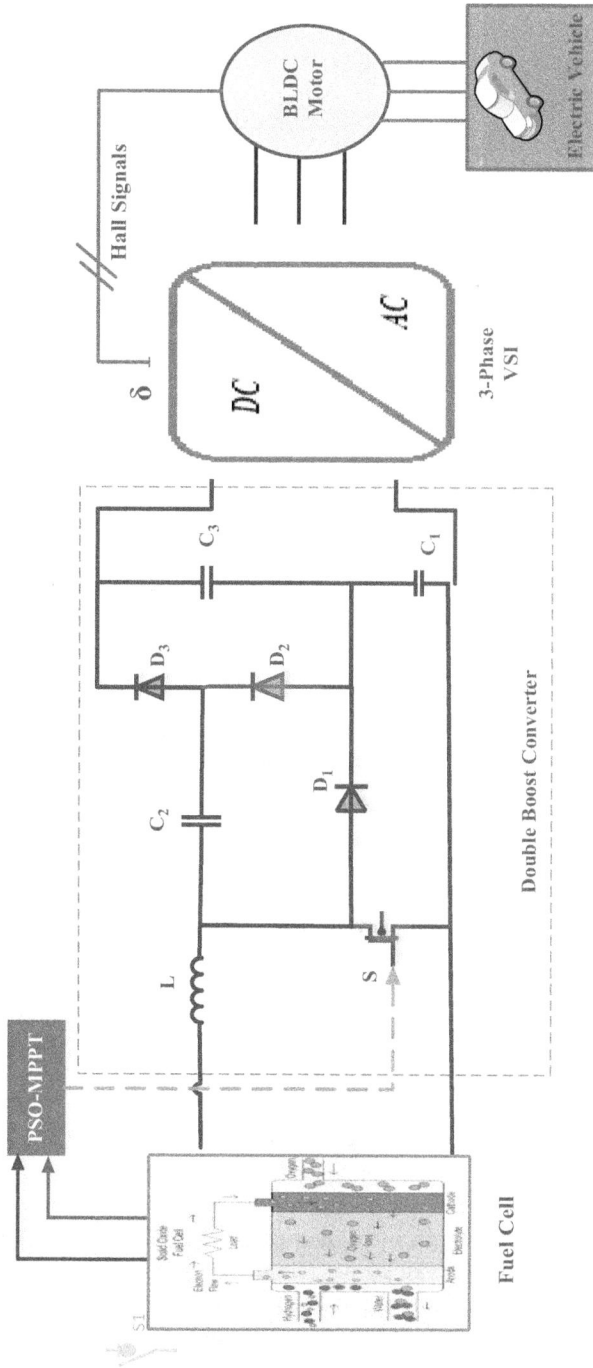

Figure 4.1 Primitive model of the proposed system.

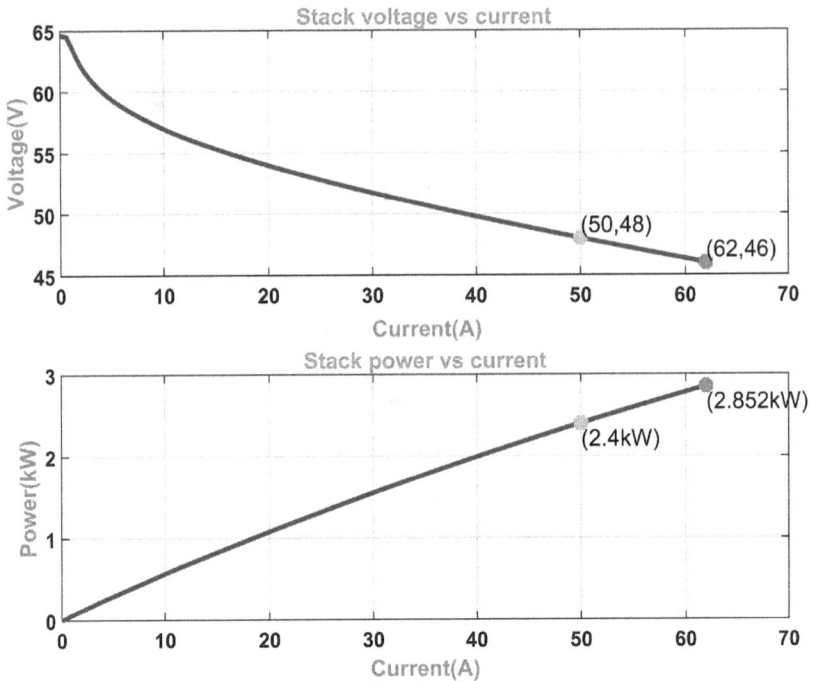

Figure 4.2 Fuel cell characteristics.

H – hydrogen; e – electron; O – oxygen; E_{cell} – generated voltage; E_O – standard potential voltage; R – gas constant (universal); T – temperature (absolute); F – Faradays constant; PH_2 – partial pressure of water; PH_2O – partial pressure of oxygen.

In order to analyse and evaluate the performance of Enriched Quadratic Boost Converter (EQBC) with different MPPT (P&O and NN-EPP) algorithms, the proposed 1.26 kW PEMFC-fed EV system is designed in a MATLAB/Simulink model. The parameter specifications of 1.26 kW PEMFC system are listed in Table 4.1.

4.2.2 Double boost converter

The proposed non-isolated single-switch high-step-up DC–DC converter is depicted in Figure 4.3. It consists of one inductor (L), three power diodes $(D_1–D_3)$, three capacitors $(C_1–C_3)$, and one power switch (S).

The circuit's operation can be explained using the four modes provided. The operational modes of the converter in the Continuous Current Mode (CCM) are shown in Figures 4.4(a) and (b). First mode the switch S is turned on in this mode of operation. D_3 is in conduction mode, while D_1, D_2, and D_0 are in the blocked state. During this action, the energy will be stored in both inductors. The capacitor C_1 is charged by the input voltage, while the

Table 4.1 Fuel cell nominal parameters

Fuel cell nominal parameters	
Stack power	
Nominal (W)	2400
Maximal (W)	2852
Fuel cell resistance (ohm)	0.12038
Nearest voltage of once cell (En) (V)	1.1576
Nominal utilisation	
Hydrogen (H_2) (%)	99.42
Oxidant (O_2) (%)	23.27
Nominal consumption	
Fuel (slpm)	23.69
Air (slpm)	56.39
Exchange current (i_0) (A)	0.83049
Exchange coefficient	0.38352
Fuel cell signal variation parameters	
Fuel cell consumption [x_H_2] (%)	99.95
Oxidant consumption [y_O_2] (%)	21
Fuel flow rate [FuelFr] at nominal hydrogen utilization	
Nominal (lpm)	4.918
Maximum (lpm)	6.098
Air flow at nominal oxidant utilisation	
Nominal (lpm)	300
Maximum (lpm)	372
System temperature (K)	338
Fuel supply pressure (bar)	6
Air supply pressure (bar)	1

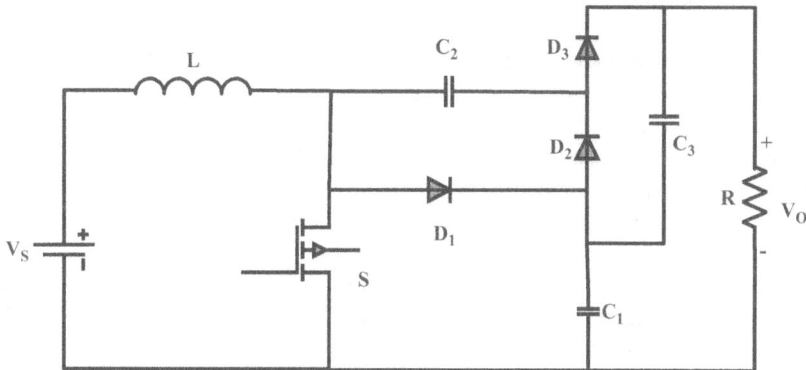

Figure 4.3 Double boost converter.

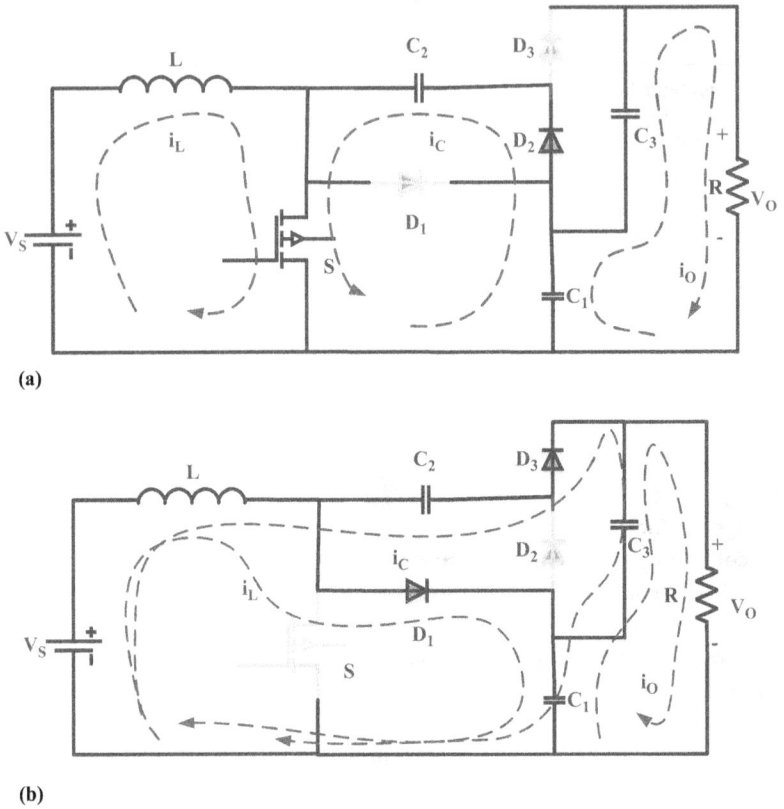

Figure 4.4 (a) ON condition. (b) OFF condition.

capacitor C_3 is charged by the input voltage. Charge will be applied to the input capacitor C_m. At this point, the output capacitor C_0 is discharged, allowing the output load to receive energy. When the current in diode D_3 reaches zero, the process is complete.

The switch remains in the ON position. D_1, D_2, D_3, and D_0 will be turned off. This phase of the operation ends when the diode D_2 begins to conduct.

In this mode, the switch S is turned on, as indicated in Figure 4.4(a). The conducting mode of diode D_2 is active. The diodes D_1, D_3, and D_0 are all turned off. C_m is a capacitor that will be charged. The route diode D_2 charges the inductors L_1 and L_2 as well as the C_1 capacitor. This mode of operation ends when the current via diode D_2 reaches zero and the main switch S is turned off.

The switch S is turned off in this mode of operation, as illustrated in Figure 4.4(b). The diode D_2 is also turned off, and the current passes through diodes D_1, D_3, and D_0.

The inductors L_1 and L_2 will be drained, while capacitor C_2 will be charged.

Table 4.2 Parameter Specifications of RQBC

Specifications	Values
Input voltage, V_{in} (V)	24
Output voltage, V_0 (V)	430
Power, P_0 (kW)	1.26
Switching frequency (kHz)	20
Inductors (mH)	$L_1 = 4.69$
	$L_2 = 4.69$
Capacitors (µF)	$C_1 = 1.87$
	$C_2 = 3.75$

In addition, the input capacitor C_m will discharge the energy. When the main switch is turned ON, one period of cycle ends. The parameter specifications of the proposed system are tabulated in Table 4.2.

4.3 PROPOSED SYSTEM CONTROL TECHNIQUES

4.3.1 MPPT controller

i. P&O

The traditional MPPT approach, perturb and observe (P&O), is employed extensively because of its simplicity and low cost. The P&O control strategy considers both the direction and the amount of the perturbation step [12–14]. The direction of the perturbation is determined by the change in the input-controlled variables. The perturbation's step size is also important in determining the strategy's convergence pace. Fixed step size P&O algorithms have yet to be introduced in the most recent system in order to improve its performance [15].

The perturbation size is used to calculate the convergence speed. The P&O's convergence speed is fast when a high perturbation size is used, but the system's efficiency is low [16]. The smaller step size is used to improve system efficiency, yet the convergence speed is slow enough to keep up with the MPP. As a result, the proposed solution for solving this problem incorporates an adaptive P&O method. The P&O method can achieve faster convergence speed and great efficiency by implementing adjustable step size. The P&O approach fails to track the maximum power when there is a sudden change in the fuel cell operating point [17].

ii. PSO

Kennedy and Eberhart devised particle swarm optimization (PSO), a population-based evolutionary approach, in 1995. It was inspired by bird flocking and fish schooling simulations [18, 19]. To locate the global maxima, the PSO algorithm offers a simple principle, fast convergence, and excellent accuracy. It is also simple to implement. PSO is made up of a flock of birds, each of which is referred to as a particle. During the optimization process, these particles fly at a particular speed and discover the global optimum place. The search space is D-dimensional, and a D-dimensional vector can be used to describe the position of the i-th particle [20–24].

The PSO algorithm is applied to the MPP tracking as follow:

Step 1: Define the input data: In this step, the input data, including the cell temperature T, the oxygen partial pressure PO_2, the hydrogen partial pressure PH_2 and the membrane water content lm, characteristics of the fuel cell system and the parameters of the PSO are defined.

Step 2: Generate the initial population randomly from the respective domains. In the proposed algorithm, the decision variable is output voltage of the fuel cell.

Step 3: Evaluate the objective function for each particle: objective function in the proposed algorithm is the output power of the fuel cell.

Step 4: Determine the best position of each particle ðPiÞ and the global best position ðPgÞ and then update the velocities and positions.

Step 5: Repeat steps 3 and 4 until a termination criterion is satisfied. In this chapter, the stopping criterion is considered the number of iterations. Furthermore, if the maximal iteration number is satisfied, the algorithm is terminated. The parameters of the PSO algorithm are determined as follows: swarm size (N) ¼ 10, dimension number (D) ¼ 1, maximum iteration (k_{max}) ¼ 100, w_{max} ¼ 0.9, w_{min} ¼ 0.4, c_1 ¼ 2, c_2 ¼ 2. These parameters are obtained by trial and error method through the use of computer simulations.

In this system, the MPP tracker connects the fuel cell system to the battery. The MPP tracker contains the boost DC-DC converter, the PSO-MPPT and the PID controller. The boost converter circuit comprises of one diode, and a power MOSFET operating at 1 kHz is used as a switching device, one inductor L ¼ 5 mH and one resistance R ¼ 0.2 mU. A Nickel-Cadmium battery with rated at 50 V, 100 Ah is used for power storage. The values of PID parameters, i.e., proportional gain (KP), integral gain (KI) and derivative gain (KD), are tuned to 0.043, 0.0001 and 0.001, respectively. These parameters are obtained by a trial and error method.

The MPP tracker tunes the boost converter duty cycle to modify the operating point of the fuel cell system to maximum power (P_{max}). The transfer function of a DC/DC boost converter is found by studying its steady-state functioning as follows:

$$V_o = \text{¼} V_{FC}^* d$$

Table 4.3 Hall signals for electronic commutation of BLDC Motor

Degree, θ	Hall sensor signals			VSI switching states					
	H_1	H_2	H_3	S_1	S_2	S_3	S_4	S_5	S_6
NA	...	...	...	...	...	...	...	...	...
0°–60°	...	...	#	...	...	...	#	#	...
60°–120°	...	#	...	...	#	#	...	...	...
120°–180°	...	#	#	...	#	...	...	#	...
180°–240°	#	...	...	#	...	...	...	...	#
240°–300°	#	...	#	#	...	...	#	...	...
300°–360°	#	#	...	...	...	#	...	...	#
NA	#	#	#	...	...	...	...	...	...
Where,	NA= Not Applicable			# = 1				... = 0	

4.3.2 Hall Effect controller for VSI

i. Voltage source inverter commutation

The commutation of the voltage source inverter (VSI) in a fuel cell-fed electric vehicle system is done by generating the switching pulses (S_1 to S_6) with the help of three hall sensor signals with the time interval of 60° each, the detailed switching pulse generation process is presented in Table 4.3 at different intervals.

4.4 RESULT AND DISCUSSION

As per the designed parameter deliberated in the previous section, the proposed FC-fed electric vehicle system is implemented in the MATLAB/Simulink model. The developed system is analyzed for duration of 1.4 sec. by considering the temperature conditions in three distinct levels of 328 Kelvin for the period of 0 to 0.4 sec., 338 Kelvin for period of 0.5 to 0.9 sec. and 318 Kelvin for the period of 1 to 1.4 sec., as presented in Figure 4.5.

Figure 4.6 gives the graphical output of FC system output voltage, current and power as per consideration of available FC temperature conditions.

Figure 4.7 gives the graphical output of FC system with a double boost converter output voltage, current and power as per the consideration of available FC temperature conditions with the implemented MPPT controllers of P&O and PSO.

The developed FC-fed electric vehicle system with a double boost converter gives output voltage, current and power of 398 V, 5.02 A and 1998 W, respectively, with P&O MPPT and, similarly, 419 V, 5.56 A and 2330 W, respectively, from PSO MPPT in span I. In span II, the output voltage, current and power are 418 V, 5.502 A and 2300 W, respectively, with P&O

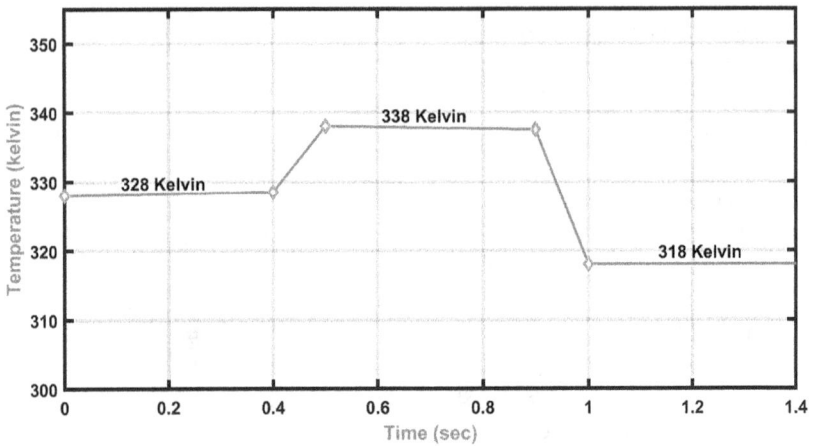

Figure 4.5 Variation of FC input temperature.

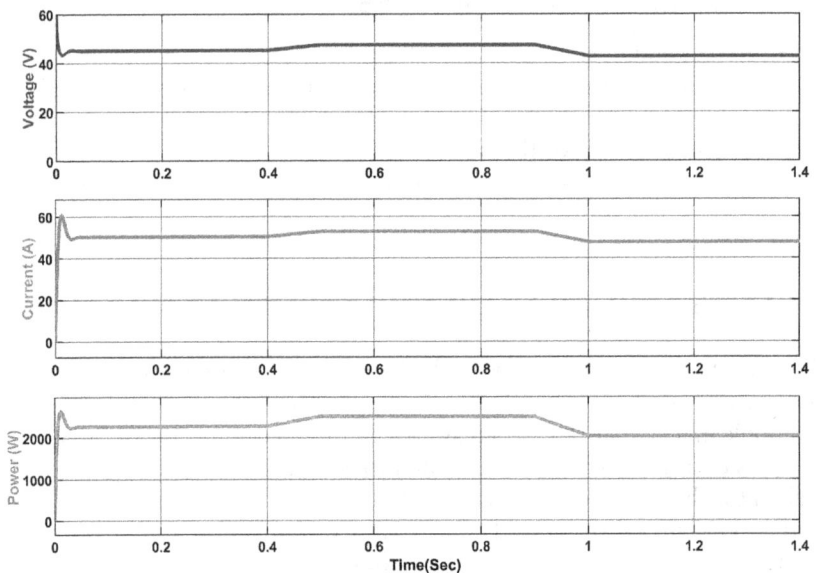

Figure 4.6 FC system output voltage, current and power at different spans.

MPPT and, similarly, 432 V, 5.51 A and 2380 W, respectively, from PSO MPPT. Similarly, in span III there is output voltage, current and power of 398 V, 5.02 A and 1998 W respectively with P&O MPPT and similarly 382 V, 4.712 A and 1800 W respectively from PSO MPPT. The proposed system with PSO MPPT controller has been given better results of 2380 W power with the considered temperature of 338 Kelvin in span II, compared to the conventional P&O MPPT method, as tabulated in Table 4.4.

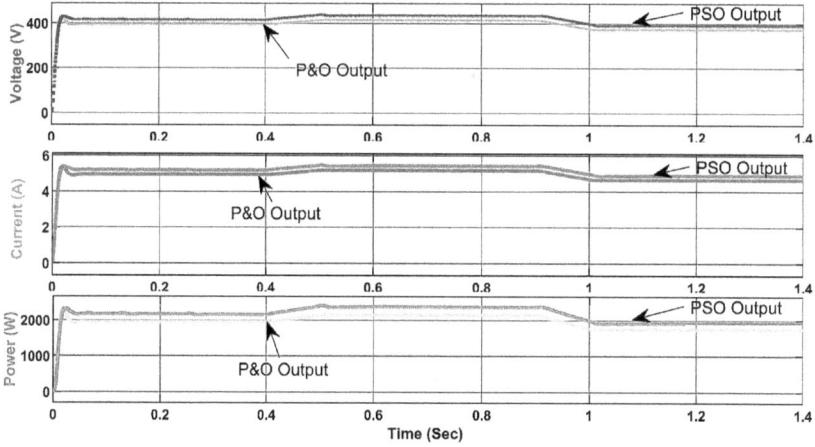

Figure 4.7 DB converter DC link voltage, current and power at different spans.

Table 4.4 System performance comparison

	Span I (0 to 0.4) sec.		Span II (0.5 to 0.9) sec.		Span III (1 to 1.4) sec.	
	P&O	PSO	P&O	PSO	P&O	PSO
Voltage (V)	398	419	418	432	398	382
Current (A)	5.02	5.56	5.502	5.51	5.02	4.712
Power (W)	1998	2330	2300	2380	1998	1800

Figure 4.8 gives the graphical output of FC system with double boost converter output voltage, current at inverter side in all three spans of considered fuel cell temperature.

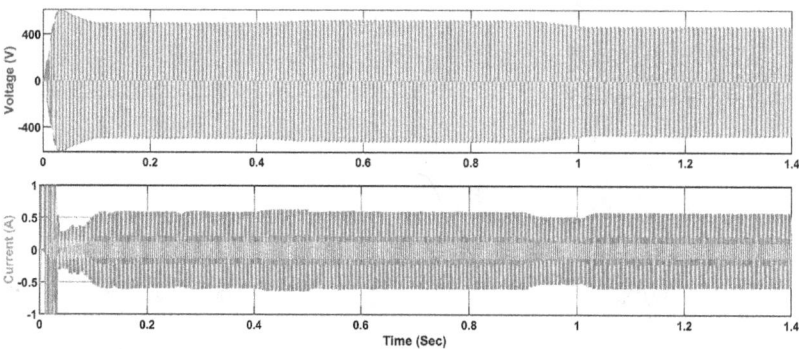

Figure 4.8 Outlined EV system output voltage and current at inverter side in all spans.

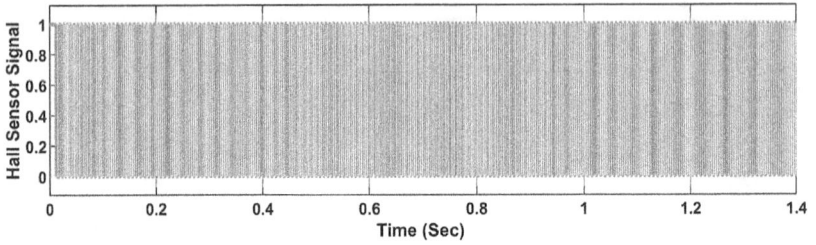

Figure 4.9 Outlined EV system Hall Sensor Signal at inverter side.

Figure 4.9 gives the graphical output of FC system hall sensor signal for the inverter control at inverter side in all three spans of considered fuel cell temperature.

Figure 4.11 gives the BLDC motor speed in rad/sec. and RPM and the BLDC torque and rotor angle of the system incorporated in the fuel cell-fed electric vehicle system with the considered fuel cell temperature conditions. From the performance analysis of fuel cell-fed electric vehicle system done in MATLAB/Simulink are presented in the simulation and result in the analysis section. The DC link voltage, current, and power values are tabulated with different MPPT techniques. The proposed system with the PSO MPPT controller has been given better results of 2380 W power with the considered temperature of 338 Kelvin in span II, compared to the conventional P&O MPPT method in MATLAB/Simulink environment.

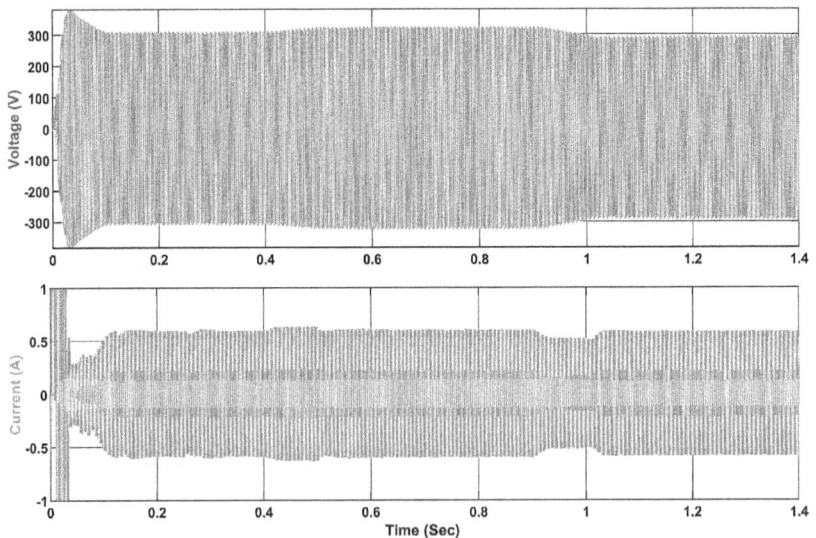

Figure 4.10 Outlined EV system voltage and current at stator side.

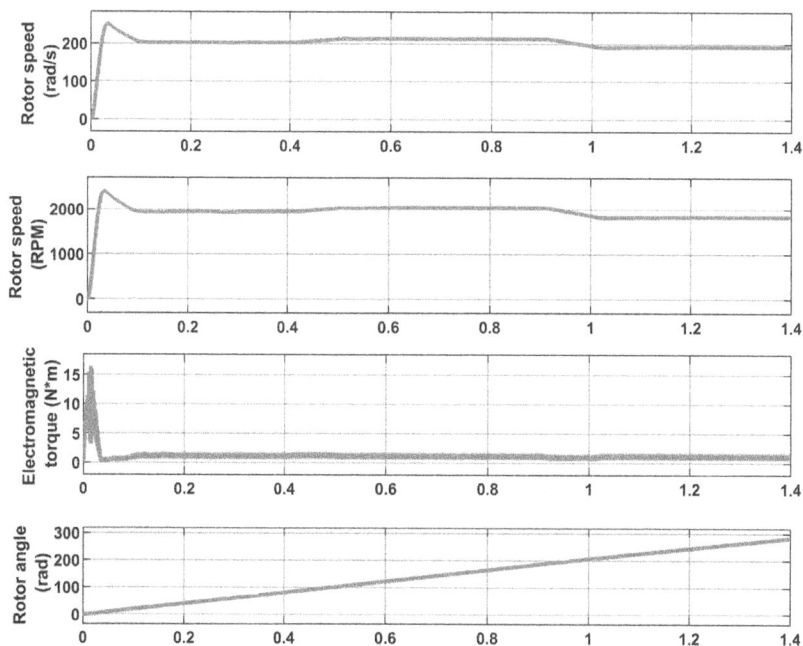

Figure 4.11 BLDC motor performance characteristics in outlined EV system.

4.5 CONCLUSION

This analysis of a fuel cell-fed electric vehicle system with a double boost converter has been done in the MATLAB/Simulink environment. The double boost converter is considered for the integration of fuel cell and EV system arrangement due to its efficient and maximum voltage transfer gain with less voltage stress across the switches compared to the conventional Boost converter. The performance is analyzed by considering the FC temperature from 318 Kelvin to 338 Kelvin in different time spans with different MPPT control techniques. The PSO MPPT method is implemented in the proposed system, the obtained results are compared to the conventional P&O MPPT method and the results are tabulated. In the EV system, BLDC motor performance is analyzed and presented in three different spans with the considered FC temperature values. The proposed system with a PSO MPPT controller has been given better results of 2380 W power with the considered temperature of 338 Kelvin in span II, compared to the conventional P&O MPPT method in the MATLAB/Simulink environment.

REFERENCES

1 Na, Woonki, Taesik Park, Taehyung Kim, and Sangshin Kwak. "Light fuel-cell hybrid electric vehicles based on predictive controllers." *IEEE Transactions on Vehicular Technology* 60, no. 1 (2010): 89–97.

2 Pany, Premananda, R. K. Singh, and R. K. Tripathi. "Active load current sharing in fuel cell and battery fed DC motor drive for electric vehicle application." *Energy Conversion and Management* 122 (2016): 195–206.

3 Charles, Rahul, and J. S. Savier. "Bidirectional DC–DC converter fed BLDC motor in electric vehicle." In *2021 International Conference on Advances in Electrical, Computing, Communication and Sustainable Technologies (ICAECT)*, pp. 1–6. IEEE, 2021.

4 Kumar, Dileep, and R. A. Gupta. "A comprehensive review on BLDC motor and its control techniques." *International Journal of Power Electronics* 14, no. 3 (2021): 292–335.

5 Hasan, Mohammad Kamrul, Md Mahmud, AKM Ahasan Habib, S. M. A. Motakabber, and Shayla Islam. "Review of electric vehicle energy storage and management system: Standards, issues, and challenges." *Journal of Energy Storage* 41 (2021): 102940.

6 Song, Ke, Yimin Wang, Cancan An, Hongjie Xu, and Yuhang Ding. "Design and validation of energy management strategy for extended-range fuel cell electric vehicle using bond graph method." *Energies* 14, no. 2 (2021): 380.

7 Verma, Shrey, Shubham Mishra, Ambar Gaur, Subhankar Chowdhury, Subhashree Mohapatra, Gaurav Dwivedi, and Puneet Verma. "A comprehensive review on energy storage in hybrid electric vehicle." *Journal of Traffic and Transportation Engineering (English Edition)* 8, no. 5 2021: 621–637.

8 Kumar, JSV Siva, and P. Mallikarjuna Rao. "Performance analysis of PID controller and sliding mode control for electric vehicle applications in interleaved double boost converter." In *Intelligent Computing in Control and Communication*, pp. 47–59. Springer, Singapore, 2021.

9 Naik, Banavath Shiva, Yellasiri Suresh, Kancharapu Aditya, and Bhukya Nageswar Rao. "A novel nine-level boost inverter with a low component count for electric vehicle applications." *International Transactions on Electrical Energy Systems* (2021): e13172.

10 Suresh, Karthik, C. Bharatiraja, N. Chellammal, Mohd Tariq, Ripon K. Chakrabortty, Michael J. Ryan, and Basem Alamri. "A multifunctional non-isolated dual input-dual output converter for electric vehicle applications." *IEEE Access* 9 (2021): 64445–64460.

11 Srinivasan, Suresh, Ramji Tiwari, Murugaperumal Krishnamoorthy, M. Padma Lalitha, and K. Kalyan Raj. "Neural network based MPPT control with reconfigured quadratic boost converter for fuel cell application." *International Journal of Hydrogen Energy* 46, no. 9 (2021): 6709–6719.

12 Habib, AKM Ahasan, Mohammad Kamrul Hasan, Md Mahmud, S. M. A. Motakabber, Muhammad I. Ibrahimya, and Shayla Islam. "A review: Energy storage system and balancing circuits for electric vehicle application." *IET Power Electronics* 14, no. 1 (2021): 1–13.

13 Chakravarthi, BN Ch V., P. Naveen, S. Pragaspathy, and VSN Narasimha Raju. "Performance of induction motor with hybrid multi level inverter for electric vehicles." In *2021 International Conference on Artificial Intelligence and Smart Systems (ICAIS)*, pp. 1474–1478. IEEE, 2021.

14 Kumar, K., Ramji Tiwari, P. Venkata Varaprasad, Challa Babu, and K. Jyotheeswara Reddy. "Performance evaluation of fuel cell fed electric vehicle system with reconfigured quadratic boost converter." *International Journal of Hydrogen Energy* 46, no. 11 (2021): 8167–8178.

15 Kumar, K., Ramji Tiwari, N. Ramesh Babu, Sanjeevikumar Padmanaban, Mahajan Sagar Bhaskar, and Vigna K. Ramachandaramurthy. "Analysis of high voltage-gain hybrid DC–DC power converter with RBFN based MPPT for renewable photovoltaic applications." In *2017 IEEE Conference on Energy Conversion (CENCON)*, pp. 294–299. IEEE, 2017.

16 Chouder, A., F. Guijoan, and S. Silvestre. "Simulation of fuzzy-based MPP tracker and performance comparison with perturb & observe method." *Journal of Renewable Energies* 11, no. 4 (2008): 577–586.

17 Rasappan, Senthilkumar, and K. Indira Devi. "Matrix converter supported by hybrid vehicle system with perturb & observe algorithm for V2G operation." *International Journal of Vehicle Structures & Systems (IJVSS)* 19, no. 3 (2020): 251–255.

18 Khouadjia, Mostepha R., Briseida Sarasola, Enrique Alba, Laetitia Jourdan, and El-Ghazali Talbi. "A comparative study between dynamic adapted PSO and VNS for the vehicle routing problem with dynamic requests." *Applied Soft Computing* 12, no. 4 (2012): 1426–1439.

19 Hannan, M. A., Mahmuda Akhtar, R. A. Begum, H. Basri, A. Hussain, and Edgar Scavino. "Capacitated vehicle-routing problem model for scheduled solid waste collection and route optimization using PSO algorithm." *Waste Management* 71 (2018): 31–41.

20 Kao, Yucheng, Ming-Hsien Chen, and Yi-Ting Huang. "A hybrid algorithm based on ACO and PSO for capacitated vehicle routing problems." *Mathematical Problems in Engineering*, 2012 (2012): 1–9.

21 Liu, Zifa, Wei Zhang, and Zeli Wang. "Optimal planning of charging station for electric vehicle based on quantum PSO algorithm." *Zhongguo Dianji Gongcheng Xuebao (Proceedings of the Chinese Society of Electrical Engineering)* 32, no. 22 (2012): pp. 39–45. Chinese Society for Electrical Engineering.

22 Guo, Xinghai, Mingjun Ji, Ziwei Zhao, Dusu Wen, and Weidan Zhang. "Global path planning and multi-objective path control for unmanned surface vehicle based on modified particle swarm optimization (PSO) algorithm." *Ocean Engineering* 216 (2020): 107693.

23 Singh, Shakti, Prachi Chauhan, and NirbhowJap Singh. "Capacity optimization of grid connected solar/fuel cell energy system using hybrid ABC-PSO algorithm." *International Journal of Hydrogen Energy* 45, no. 16 (2020): 10070–10088.

24 Ahmadi, S., S. Abdi, and M. Kakavand. "Maximum power point tracking of a proton exchange membrane fuel cell system using PSO-PID controller." *International Journal of Hydrogen Energy*, 42, no. 32 (2017): 20430–20443.

Chapter 5

Structural, finite element and simulation analysis for wireless power transfer of power pad for electric vehicles

Bilal Alam, Mohd Tariq, Safwan Mustafa, Wajid Ali, Azam Khan and Khaliqur Rahman
ZHCET, Aligarh Muslim University, Aligarh, India

Mohammed A. Bou-Rabee
College of Technical Studies, PAAET, Safat, Kuwait

CONTENTS

5.1 INTRODUCTION

The emerging electric vehicle (EV) industry has boosted the demand for ergonomic and effective battery charging techniques. Researchers have demonstrated significant progress in implementing wireless charging systems for electric vehicles and enhancing their performance [1–5]. The plug-in method of charging EVs is not convenient, and in wet environments, it might cause safety issues. In addition, both tripping, and leakage caused by a damaged aged connection may result from charging cables on the floor. EV owners have promoted Wireless Power Transfer (WPT) technology as this eliminates all of the troublesome charging issues. The WPT methodology does not necessitate any physical interaction between cars and charging sources so that the inconveniences and hassle caused by a typical conducting methodology can be addressed [6, 7]. Charging becomes the simplest chore by wirelessly transmitting energy to the EV. WPT processes transmit energy via photonic light waves, magnetic or electric effects, or both. WPT has attracted a lot of interest in the power electronics sector because of its potential advantages in terms of convenience, versatility, reliability, and security [8–12]. In the present era, the WPT technology has been adopted effectively in biomedical implant applications, unmanned aerial vehicles (UAV), electric vehicles (EV), underwater power supply, etc. in the electric vehicle WPT system, the design, size, and position of the magnetic core and coil windings have a detrimental impact on wireless charging performance and transmission efficiency.

In practice, lateral misalignment across the transmitter and receiver coils is highly prevalent. This has proved to be a significant obstacle in WPT development since the coil's configurations have a substantial influence on mutual inductance [13, 14]. Misalignment causes a decline in mutual inductance, leading to instability and lowering the efficiency of the transfer. Different approaches have been suggested to optimize the misalignment of the WPT system focused on topologies, control strategies, and coil designs. Another alternative to reducing the misalignment impact is to develop new coil structures. Many magnetic couplers with novel configurations have been proposed to enhance misalignment adaptability [15, 16].

The power pad and coil arrangement for electric vehicles (EV) wireless charging architecture are the most important concerns with regard to high-frequency power converters, electromagnetic field protection, metal, and foreign object detection [17–19]. Enhancing the quality factor and geometrical structure is the main concern in power pad design. Wireless power transmission efficiency is proportional to the coupling coefficient and the performance of the transmitter and receiver. The static wireless charging systems' primary coil structures are unipolar, bipolar, and solenoid coil structures. Unipolar structures yield vertical magnetic flux and have an elevated coupling coefficient [6]. Bipolar coil topologies, by contrast, generate horizontal magnetic flux, resulting in a lower coupling coefficient. Alternative classifications for the power pad include the non-polarized structure of the coil, for instance, circular pads, rectangular pads, and square power pads; on the other hand, there are polarized power pads such as DD, DDQ, and solenoid. Non-polarized pads require a separate pole and the flux is dispersed in all dimensions. In the polarized power pad, two opposing poles, north, and south, are formed.

The magnetic core influence on the wireless power pad configuration is analyzed in Section 5.2 of this manuscript. Section 5.3 depicts the proposed work's system modelling. Section 5.4 focuses on the Finite Element Analysis of several kinds of wireless power pads and Section 5.5 discusses the conclusion.

5.2 MAGNETIC CIRCUIT ANALYSIS FOR RECTANGULAR, DD, DDQ COIL SYSTEM

To analyze the influence of the wireless power pad's core, a theoretical magnetic study of D coil, DD coil, and DDQ coil is conducted. The magnetic flux produced by the coil is released on all sides. In comparison to the core material, the flux has a relatively elevated air reluctance. Thus, the flux generated in the core can be overlooked in contexts of flux through the air. The advantages of these coils have been proposed by Budhia et al. and Zaheer et al. and many more researchers [20–22]. The DD coil appears to be the most powerful and effective. We investigated three kinds of coil structures, D, DD, and DDQ, to compare and observe the core's influence on these three coil structures. The following sections discuss some of the assumptions made during this analysis.

- The coil pad's magnetic reluctance is ignored. Because air has a relatively high reluctance in comparison to such a magnetic core.
- The magnetic field generated at the coil's sides is overlooked.
- The system under consideration is perfectly aligned as well as symmetrical.

The magnetic induction method is used for WPT, with the air gap acting as the magnetic coupling medium. There are two distinct flux paths between the transmission and reception: one within the magnetic core and one through the air. Because the reluctance through the air is greater than the magnetic reluctance within the magnet, it may be overlooked. The standardization of the magnetic field level has been worked by SAE 2954.

5.2.1 Rectangular coil pad structure

A rectangular coil (or D coil) provides a larger flux area and less flux leakage along the edges. Usually, this is beneficial in creating track coils for dynamic wireless charging as it has a higher tolerance for lateral alignment. D coils are simple to construct, lightweight and require less space. D coils are more cost-effective and have a great capacity for transmitting efficiency. Figure 5.1(a) shows the D coil structure. The equation is derived in [23]. Due to the symmetrical construction, mutual fluxes are identical on both sides. The coupling coefficient (k) is deduced in [23], and this is as follows:

$$k = \frac{\text{Self reluctance of Coil1}}{\left[\text{Self reluctance of coil1} + \left(2 \times \text{mutual magnetic reluctance}\right)\right]} \quad (5.1)$$

(a) Rectangular Coil Pad   (b) DD Coil Pad

(c) DDQ coil Structure

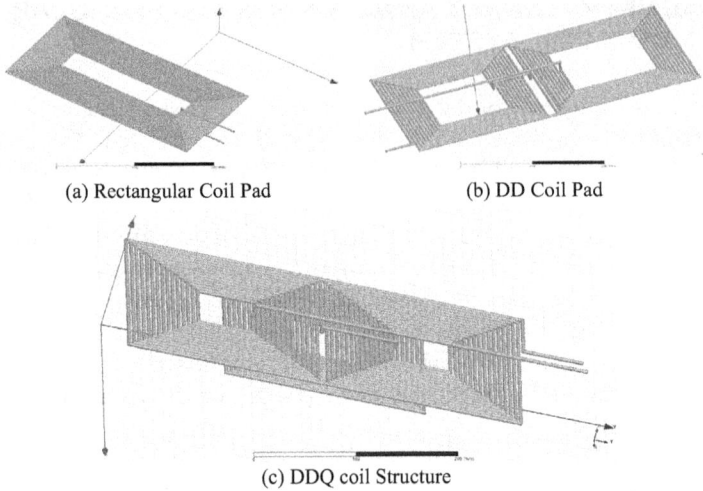

Figure 5.1 Structure of (a) rectangular coil, (b) DD and (c) DDQ coil pad.

From Equation (5.1) it can be seen that the coupling coefficient is dependent on the core materials and the proper placement. In D coil k is inversely proportional to the mutual reluctance.

5.2.2 DD coil pad structure

The DD coil has a high coupling coefficient and a low lateral misalignment tolerance. Figure 5.2(b) delineates the systemic diagram of the DD coil pad. The DD coil is comprised of two coils, which are attached in series. Each coil has electromotive forces (say F_1 and F_2). The detailed derivation is derived in [23]. The k of the DD coil system is as follows.

$$k = \frac{\left(\dfrac{F_1 + F_2}{R_{m1}}\right) + \left(\dfrac{F_1 + F_2}{2R_{m2}}\right)}{\dfrac{F_1}{R_{s3}} + \dfrac{F_1 + F_2}{R_{s1}} + \dfrac{F_2}{R_{s3}} + \dfrac{F_1 + F_2}{R_{m1}} + \dfrac{F_1 + F_2}{R_{m1}} + \dfrac{F_1 + F_2}{2R_{m2}}} \qquad (5.2)$$

R_{s1}, R_{s2}, R_{s3} are the self-reluctance, and R_{m1}, R_{m2} are the mutual reluctance of the coils. From Equation (5.2), we can say that the K depends on the core material's positioning. By increasing or decreasing the R_{m1} and R_{m2}, we can control the coupling coefficient (k).

5.2.3 DDQ coil structure

The DDQ coil is the advanced version of the DD coil pad. In the DD coil, there is an extra coil pad that is placed over the DD pad. This additional pad

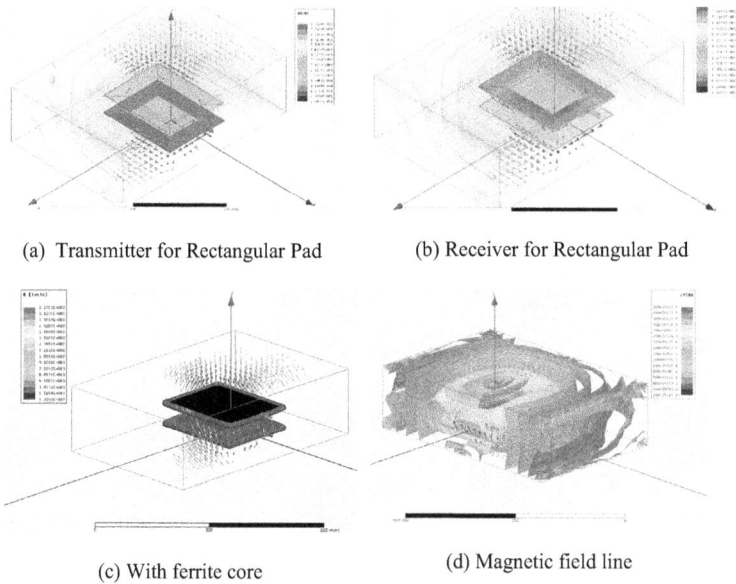

(a) Transmitter for Rectangular Pad

(b) Receiver for Rectangular Pad

(c) With ferrite core

(d) Magnetic field line

Figure 5.2 A comparative simulation results for rectangular coil.

did not produce any electromagnetic field. The additional pad is isolated from the DD pad. DDQ generated the field's vertical and horizontal flux components, since DDQ has a greater tendency to bear more misalignment. Figure 5.1(c) shows the diagram of the DDQ pad.

5.3 SYSTEM MODELLING AND ANALYSIS

Figure 5.3 illustrates the suggested WPT system's circuit diagram. There is a direct current source, an inverter that converts the direct current to alternating current, an SP topology, a full bridge rectifier, a magnetic coupler, and a load in this system. The analysis is performed using the first harmonic

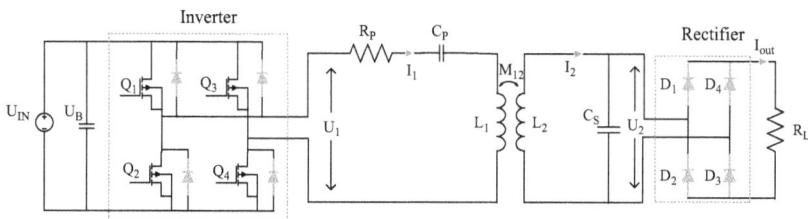

Figure 5.3 Circuit diagram of the proposed WPT.

approximation (FHA) algorithm. The fundamental component of the output voltage of the system is given as follows;

$$U_1 = \frac{4}{\pi} U_{IN} \sin(\omega t) \tag{5.3}$$

Where U_{IN} denotes the total system input voltage, ω is the angular frequency, equivalent to the product of $2\pi f$. The phasor for the output voltage (U_1) of the inverter is given by

$$U_1 = \frac{2\sqrt{2}}{\pi} U_{IN} < 0 \tag{5.4}$$

Where L_1 and L_2 are the self-inductor, and C_P and C_S are the transmitter's and receiver's capacitors, respectively. R_P is the resistor in the primary of the transmitter where R_L is the load of the secondary of the receiver. M_{12} symbolizes the mutual inductance of the transmission and reception. We use Series-Parallel (SP) compensation topology for the grounds of simplicity, since SP is more economically suitable for higher transmission. Because resonance is ensured, it is also best adapted for varying load circumstances. In this system, we run our system on a resonant angular frequency and it is given by

$$\omega = \frac{1}{\sqrt{L_1 \times C_P}} = \frac{1}{\sqrt{L_2 \times C_S}} \tag{5.5}$$

Using Kirchhoff's law, we can obtain the circuit equation;

$$U_1 = I_1 \left(R_P + \frac{1}{j\omega C_P} + j\omega L_1 \right) - j\omega M_{12} I_2 \tag{5.6}$$

$$U_1 = I_1 R_P - j\omega M_{12} I_2$$

From Equation (5.5)

$$0 = -j\omega M_{12} I_1 + (C_2 \| R_L) I_2 \tag{5.7}$$

$$0 = -j\omega M_{12} I_1 + \left(\frac{R_L}{j\omega C_2 R_L + 1} \right) I_2 \tag{5.8}$$

Therefore, the required input and output current will be

$$I_1 = \frac{R_L U_1}{R_L R_P - (j\omega M_{12})^2 (j\omega C_2 R_L + 1)} \tag{5.9}$$

$$I_2 = \frac{U_1 (j\omega M_{12})(j\omega C_2 R_L + 1)}{R_L R_P - (j\omega M_{12})^2 (j\omega C_2 R_L + 1)} \tag{5.10}$$

From Equation (5.9), the Load (R_L), Capacitance C_2, and M_{12} will be

$$R_L = \frac{I_1 (j\omega M_{12})^2}{(U_1 - I_1 R_P)(1 - j\omega C_2)(j\omega M_{12})^2} \tag{5.11}$$

$$C_2 = \frac{I_1 R_L R_P + I_1 (j\omega M_{12})^2 - R_L U_1}{I_1 R_L (j\omega M_{12})^2} \tag{5.12}$$

$$M_{12} = \sqrt{\frac{\dfrac{R_L R_P I_1 - R_L U_1}{I_1 (j\omega C_2 R_L + 1)}}{j\omega}} \tag{5.13}$$

5.4 FINITE ELEMENT ANALYSIS AND SIMULATION RESULT

Since air acts as the magnetic flux path for wireless power transfer, the analytical technique for calculating mutual inductance is complex and challenging. To obtain the most accurate mutual inductance value a FEA analysis is used. Ansys Maxwell is an electromagnetic field solver for electric machines, wireless charging, transformer, etc. Ansys Maxwell is utilized to obtain optimal performance while using the least amount of processing power. The magnetostatic analysis performed in Ansys Maxwell measured the mutual inductance, self-inductance, and coupling coefficient. Once we've determined the self- as well as the mutual inductance values for each coil, we can proceed. The next stage is to perform a transient analysis simulation with Ansys Simplorer to examine the phase shift of the current, voltage across the load, coil, and components.

The initial step in the simulation procedure was to simulate the coil using the coil characteristics provided in Table 5.1. The next step was to create a symmetrical design with independent stimulation for every coil. This is done to validate the analytical observation.

Table 5.1 Specification of coils for FEA analysis

WPT pad	Distance between receiver and transmitter (mm)	Diameter of wire (mm)	Material	Size of pad (mm)	Number of turns	Size of ferrite (mm)
Rectangular pad	10	4	Copper	400 × 200	15 in Tx 15 in Rx	400 × 200
DD pad	10	4	Copper	200 × 200	15 in Tx 15 in Rx	400 × 200
DDQ pad	10	4	Copper	400 × 200 (Rx and Tx) 200 × 200 (Middle)	15 in Tx 15 in Rx 15 in Middle	400 × 200

5.4.1 Simulation result obtained by Ansys Maxwell

The mutual inductance of the rectangular coil pad structure is then verified and determined using FEA. The rectangular coil pad contains numerous advantages [6, 24–26], including coupling capability with other coil structures, greater misalignment acceptance, and elevated coupling coefficient. Figure 5.2(a) and (b) represents the transmitter and receiver of the rectangular coil pad. Flux lines aligned and became directional once the ferrite core was placed across the transmitting and receiving coils. As illustrated in Figure 5.2(c) and (d), the leakage flux was limited by the ferrite core and concentrated at the centre.

The magnetic field structure of the DD coil is much more sophisticated. According to a theoretical study, the use of a ferrite core would improve flux linking and the coupling coefficient. As per Equation (5.2), the coupling coefficient is affected by both reduced self-reluctance and increased mutual reluctance. Figures 5.4(a) and 5.4(b), despite the transmitter, is rectangular and the receiver is the DD coil pad. On Ansys Maxwell, the DD coil has been analyzed with the placement of the ferrite core as shown in Figures 5.4(c) and 5.4(d). Figure 5.5 shows the magnetic-flux-produced DDQ coil structure. The field is concentrated in the middle; there is no flux coming out of the core.

Table 5.2 shows the individual inductance, mutual inductance, magnetic flux coupling coefficient, and magnetic flux. In Table 5.2, the bold font indicates the maximum coupling coefficient in rectangular (RR) pads where the receiver and transmitter are both rectangular. The bold and italics font in the similar table shows the minimum coupling coefficient when the receiver and transmitter are both in DD shape.

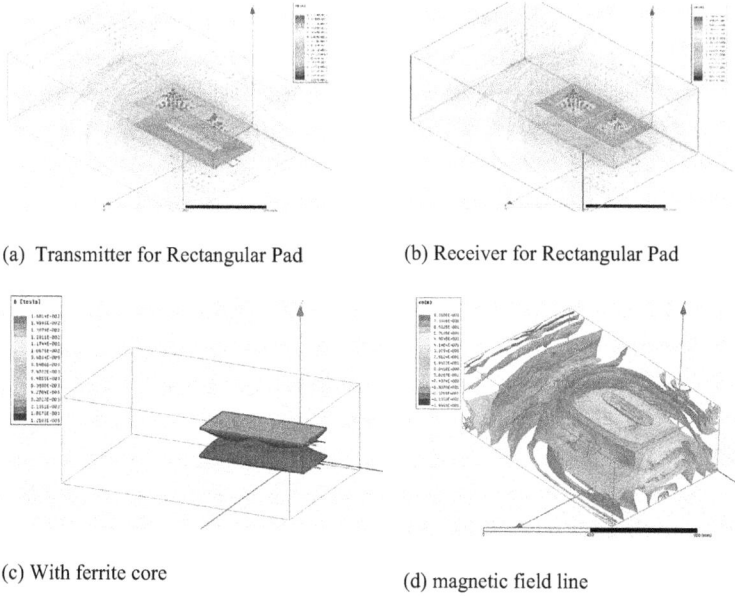

(a) Transmitter for Rectangular Pad

(b) Receiver for Rectangular Pad

(c) With ferrite core

(d) magnetic field line

Figure 5.4 A comparative simulation results for DD coil.

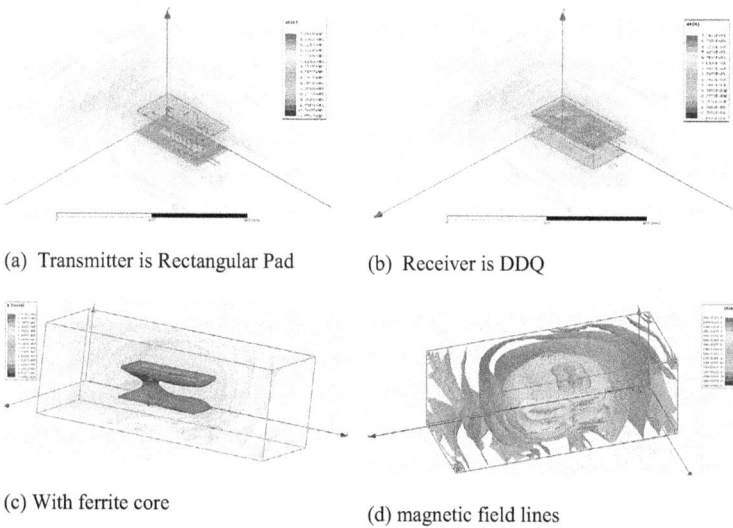

(a) Transmitter is Rectangular Pad

(b) Receiver is DDQ

(c) With ferrite core

(d) magnetic field lines

Figure 5.5 Results obtained by Ansys Maxwell for DDQ coil.

Figure 5.6 shows the graphical representation of the coupling coefficient. By this figure, we can observe that the best coupling coefficient value is given when the transmitter and receiver are both rectangular. In Figure 5.7, the dark grey colour of the bar represents the magnetic flux in the transmitter and the light grey shows the magnetic flux in the receiver.

Table 5.2 Results obtained by Ansys Maxwell

Types of WPT pads	LI (μH)	Mutual inductance (μH)	L2 (μH)	Magnetic flux (Tx) (Wb)	Coupling coefficient	Magnetic flux (Rx) (Wb)
D-DD	68.2092	15.6677	117.8493	0.000682	0.174	0.001022
D-DDQ	73.5499	15.00460	96.2253	0.000886	0.178	0.001112
DD-DD	*98.29009*	*15.997560*	*98.2874*	*0.001143*	*0.162*	*0.001143*
Rectangular	**42.6830**	**12.3268**	**41.7629**	**0.000600**	**0.2103**	**0.000591**

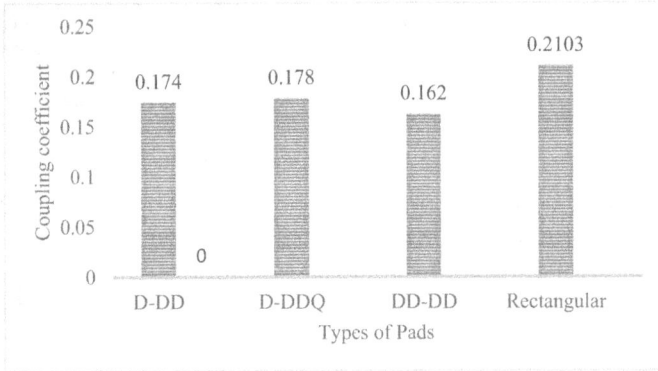

Figure 5.6 Coupling coefficient comparison.

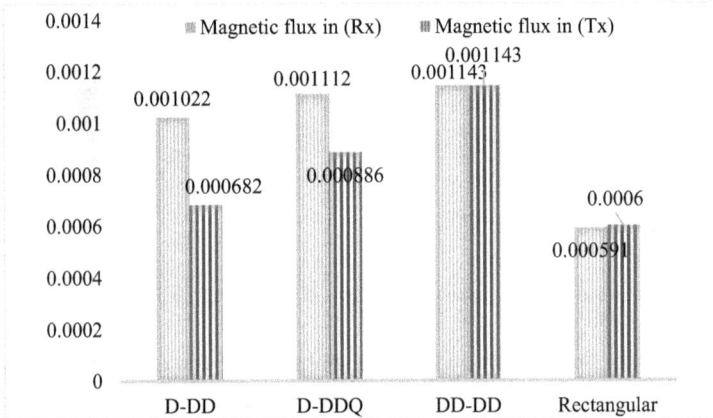

Figure 5.7 Comparative results are shown graphically.

5.5 CONCLUSION

The 3D Maxwell simulation is performed to analyze and authenticate the core's role in developing the WPT charging power pad. Rectangular, DD, and DDQ coils are explored and contrasted, based on magnetic flux and the coupling coefficient. The DD coil pad's magnetic field pattern is most precisely aligned. The ferrite core is placed in the coil's most optimal zone, and the influence of ferrite on the coupling is demonstrated in this simulation. As the ferrite materials confine the magnetic fields, no harm comes to the nearby objects, and non-ferrite regions do not interact with each other. This research will influence the field pattern produced by the mentioned configuration and the ferrite core's influence.

ACKNOWLEDGEMENTS

This research was funded by the collaborative research grant scheme (CRGS) project, Hardware-In-the-Loop (HIL) Lab, Department of Electrical Engineering, Aligarh Muslim University, India having project numbers CRGS/MOHD TARIQ/01 and CRGS/MOHD TARIQ/02.

The authors also acknowledge the technical support provided by the Hardware-In-the-Loop (HIL) Lab, Department of Electrical Engineering, Aligarh Muslim University, India.

REFERENCES

[1] S. L. Ho, J. Wang, W. N. Fu, and M. Sun, "A comparative study between novel witricity and traditional inductive magnetic coupling in wireless charging," *IEEE Trans. Magn.*, vol. 47, no. 5, pp. 1522–1525, 2011, doi:10.1109/TMAG.2010.2091495

[2] S. S. Mohammed, K. Ramasamy, and T. Shanmuganantham, "Wireless power transmission – A next generation power transmission system," *Int. J. Comput. Appl.*, vol. 1, no. 13, pp. 102–105, 2010, doi:10.5120/274-434

[3] D. Ongayo, and M. Hanif, "An overview of single-sided and double-sided winding inductive coupling transformers for wireless Electric Vehicle charging," In *2015 IEEE 2nd Int. Futur. Energy Electron. Conf. IFEEC 2015*, 2015, doi:10.1109/IFEEC.2015.7361593

[4] F. Musavi, M. Edington, and W. Eberle, "Wireless power transfer: A survey of EV battery charging technologies," In *2012 IEEE Energy Convers. Congr. Expo. ECCE 2012*, pp. 1804–1810, 2012, doi:10.1109/ECCE.2012.6342593

[5] P. S. Subudhi, and S. Krithiga, "Wireless power transfer topologies used for static and dynamic charging of EV battery: A review," *Int. J. Emerg. Electr. Power Syst.*, vol. 21, no. 1, pp. 1–34, 2020, doi:10.1515/ijeeps-2019-0151

[6] D. Patil, M. K. McDonough, J. M. Miller, B. Fahimi, and P. T. Balsara, "Wireless power transfer for vehicular applications: Overview and challenges," *IEEE Trans. Transp. Electrif.*, vol. 4, no. 1, pp. 3–37, 2017, doi:10.1109/TTE.2017.2780627

[7] E. R. Joy, B. K. Kushwaha, G. Rituraj, and P. Kumar, "Analysis and comparison of four compensation topologies of contactless power transfer system," In *2015 4th Int. Conf. Electr. Power Energy Convers. Syst. EPECS 2015*, 2015, doi:10.1109/EPECS.2015.7368544

[8] C. Panchal, S. Stegen, and J. Lu, "Review of static and dynamic wireless electric vehicle charging system," *Eng. Sci. Technol. an Int. J.*, vol. 21, no. 5, pp. 922–937, 2018, doi:10.1016/j.jestch.2018.06.015

[9] K. Ichikawa, and H. Bondar, "Power transfer system," pp. 255–259, 2012, [Online]. Available: https://www.google.ch/patents/US20120299392

[10] M. J. Chabalko, J. Besnoff, and D. S. Ricketts, "Magnetic field enhancement in wireless power with metamaterials and magnetic resonant couplers," *IEEE Antennas Wirel. Propag. Lett.*, vol. 15, no. c, pp. 452–455, 2016, doi:10.1109/LAWP.2015.2452216

[11] J. Dai, and D. C. Ludois, "A survey of wireless power transfer and a critical comparison of inductive and capacitive coupling for small gap applications," *IEEE Trans. Power Electron.*, vol. 30, no. 11, pp. 6017–6029, 2015, doi:10.1109/TPEL.2015.2415253

[12] M. M. El Rayes, G. Nagib, and W. G. Ali Abdelaal, "A review on wireless power transfer," *Int. J. Eng. Trends Technol.*, vol. 40, no. 5, pp. 272–280, 2016, doi:10.14445/22315381/ijett-v40p244

[13] J. L. Villa, J. Sallán, J. F. Sanz Osorio, and A. Llombart, "High-misalignment tolerant compensation topology for ICPT systems," *IEEE Trans. Ind. Electron.*, vol. 59, no. 2, pp. 945–951, 2012, doi:10.1109/TIE.2011.2161055

[14] A. Ahmad, M. S. Alam, and R. Chabaan, "A comprehensive review of wireless charging technologies for electric vehicles," *IEEE Trans. Transp. Electrif.*, vol. 4, no. 1, pp. 38–63, 2017, doi:10.1109/TTE.2017.2771619

[15] A. S. Bilal Alam, Maaz Nusrat, Zeeshan Sarwer, and Mohammad Zaid, "A general review of the recently proposed asymmetrical multilevel inverter topologies," *Innovations Cyber Phys Syst. Lect Notes Electr Eng*, 2021, doi:10.1007/978-981-16-4149-7_61

[16] G. A. J. Elliott, S. Raabe, G. A. Covic, and J. T. Boys, "Multiphase pickups for large lateral tolerance contactless power-transfer systems," *IEEE Trans. Ind. Electron.*, vol. 57, no. 5, pp. 1590–1598, 2010, doi:10.1109/TIE.2009.2031184

[17] A. Kurs, A. Karalis, R. Moffatt, J. D. Joannopoulos, P. Fisher, and M. Soljačić, "Wireless power transfer via strongly coupled magnetic resonances," *Science (80-.)*., vol. 317, no. 5834, pp. 83–86, 2007, doi:10.1126/science.1143254

[18] M. Bertoluzzo, G. Buja, and H. K. Dashora, "Design of DWC system track with unequal DD coil set," *IEEE Trans. Transp. Electrif.*, vol. 3, no. 2, pp. 380–391, 2017, doi:10.1109/TTE.2016.2646740

[19] Y. Li, T. Lin, R. Mai, L. Huang, and Z. He, "Compact double-sided decoupled coils-based WPT systems for high-power applications: Analysis, design, and experimental verification," *IEEE Trans. Transp. Electrif.*, vol. 4, no. 1, pp. 64–75, 2017, doi:10.1109/TTE.2017.2745681

[20] M. Budhia, G. Covic, and J. Boys, "A new IPT magnetic coupler for electric vehicle charging systems," *IECON 2010 - 36th Annual Conference on IEEE Industrial Electronics Society*, 2010, pp. 2487–2492, doi: 10.1109/IECON.2010.5675350

[21] M. Budhia, G.A. Covic, and J.T. Boys, "No title design and optimisation of magnetic structures for lumped inductive power transfer systems," In *IEEE Energy Convers, no. Congr. Expo*, 2009, 2081–2088, 2009.

[22] M. Budhia, G.A. Covic, and J.T. Boys, "Design and optimization of circular magnetic structures for lumped inductive power transfer systems," *IEEE Trans. Power Electron.*, vol. 26, no. 11, pp. 309, 2011.

[23] A. Ahmad, M. S. Alam, R. Chabaan, and A. Mohamed, "Comparative analysis of power pad for wireless charging of electric vehicles," pp. 1–7, 2019, doi:10.4271/2019-01-0865

[24] J. M. Miller, F. Ieee, and A. Daga, "Elements of wireless power transfer essential to high power charging of heavy duty vehicles," vol. 7782, no. c, 2015, doi:10.1109/TTE.2015.2426500

[25] F. Wen, X. Chu, Q. Li, and W. Gu, "Compensation parameters optimization of wireless power transfer for electric vehicles," *Electron.*, vol. 9, no. 5, pp. 1–12, 2020, doi:10.3390/electronics9050789

[26] Z. Zhang, R. Pittini, M. A. E. Andersen, and O. C. Thomsena, "A review and design of power electronics converters for fuel cell hybrid system applications," *Energy Procedia*, vol. 20, pp. 301–310, 2012, doi:10.1016/j.egypro.2012.03.030

Chapter 6

Performance analysis and misalignment effect of power pad for dynamic wireless power charging of electrical vehicles

Bilal Alam, Wajid Ali and Mohd Tariq
ZHCET, Aligarh Muslim University, Aligarh, India

Mohammed A. Bou-Rabee
College of Technical Studies, PAAET, Safat, Kuwait

CONTENTS

6.1 INTRODUCTION

In recent years, there has been major concern expressed about the exhaustible nature of fossil fuels given the reliance of most conventional vehicles on a continuous supply. In addition, the principal reason for atmospheric pollution and problems related to the degradation of the environment is the present-day significant usage of internal combustion (IC) vehicles. Various government organizations across the world have embraced electric vehicle (EV) technology as an alternative to IC engine vehicles in order to minimize both pollution and the dependence on fossil fuels [1, 2].

EVs have been around for over a century, ever since Thomas Parker invented the first car based on the concept of an electric vehicle in 1884 [3, 4]. Due to environmental pollution and fossil fuel-related difficulties, as mentioned above, there has been a surge in electric vehicle development since the turn of the 21st century. Because electric vehicles do not pollute the environment or emit noise, and because they are highly efficient, they are an increasingly attractive proposition in the automobile sector [5, 6].

DOI: 10.1201/9781003311195-6

Due to the increased demand for wireless electric power transmission by customers, technologies connected to wireless power transmission are undergoing rapid development. The wireless transmission of electric power is a technology that allows electric power to be transmitted without the use of conducting cables [7]. Researchers based at MIT have successfully executed an experiment transferring 60 W of electric power through the air over a distance of 2 m, making wireless power transmission more appealing. Magnetic resonance coupling technology was used to transmit the electricity [8]. Following that, other experiments were carried out to increase transmission distance and power transfer efficiency. A research study at the University of Auckland has developed a permanent wireless charging station for electric vehicles that can transfer up to 5 kW of electricity. They were able to successfully transfer electric power at a distance of 20 cm with an efficiency of 90%, yielding positive outcomes. In another project, a team from the Korea Advanced Institute of Science and Technology (KAIST) implemented an On-line Electric Vehicle (OLEV) project that was able to transfer 60 kW of electric power to buses with a high-power transfer efficiency of 70% [9, 10].

The inductively coupled power transfer (ICPT) approach has enabled high-level power transmission through advanced techniques of electric vehicle charging. ICPT is well known because it can transfer high power in a variety of applications, including EVs. This enables a rapid charging process, as well as the optimization of power transmission and control over the loss caused by poor magnetic coupling, by altering the frequency. The ICPT approach can be used by both stationary and driving electric vehicles [11–13].

6.2 DYNAMIC WIRELESS CHARGING

The dynamic Wireless Power Transfer (WPT) system is the second form of charging an electric vehicle by electromagnetic induction and the vehicles can be charged while they are in motion. It is similar to static charging, but the energy transfer takes place through the magnetic resonant coupling (MRCWPT) method. The transmitter coils produce high-frequency magnetic fields and these varying fields couple with the receiver coil placed in the vehicle. As a result, power is transferred through magnetic coupling and battery charges. To achieve dynamic charging, complete roads are designed in such a way that they act as a large transmitter system, and the vehicle moving on the road (which acts as a receiver system) becomes charged while it moves along. The advantages of dynamic charging are the low stand-in charge time, and the lower depth of discharge. This leads to a longer battery life since the battery could be charged from time to time; it also allows the battery to be a smaller size. This method of charging electric vehicles is very efficient and reliable and helps to overcome range anxiety. But certain factors may reduce the efficiency of the system, such as foreign objects on the road, an abrasive road surface or changes in the coil structure of the transmitting coil. To date, the dynamic charging is still at the experimental stage since certain challenges need to be overcome: these include flux leakage limitation, the applicability of different coil types, universal coil type selection, and real-time coil misalignment estimation. As can be seen in Figure 6.1, the roads are incorporated with dynamic charging pads and the vehicle starts to be charged as it moves over the pads on the road.

6.3 POSSIBLE MISALIGNMENT

The analytical model described in this chapter is used to determine the mutual inductance and coupling coefficient, which are affected by the secondary coil alignment. As a result, there are a number of possible misalignments that can occur when we move the secondary coil:

a. Vertical misalignment (VM)
b. Planar misalignment (PM)
c. Angular misalignment (AM)
d. Horizontal misalignment (HM)
e. Planar with horizontal misalignment (PHM)
f. Angular with horizontal misalignment (AHM)

In this chapter, we focus solely on horizontal misalignment. This occurs when the secondary coil of the system moves along the x-axis. The centres of the main and secondary coils are misaligned by Δx in the x-direction in

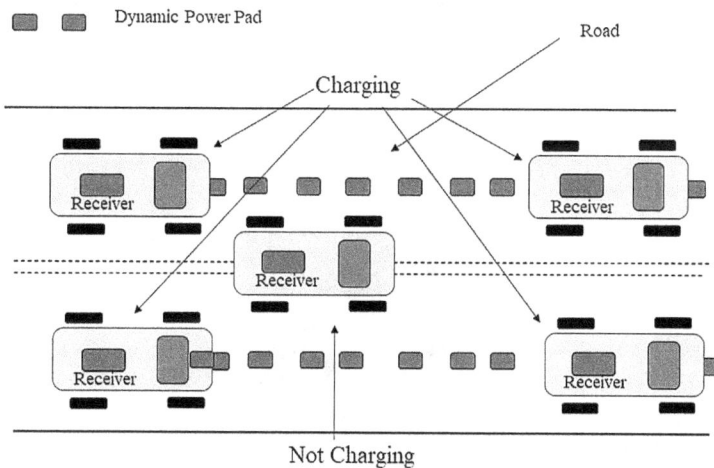

Figure 6.1 Block diagram of dynamic wireless charging system of EVs.

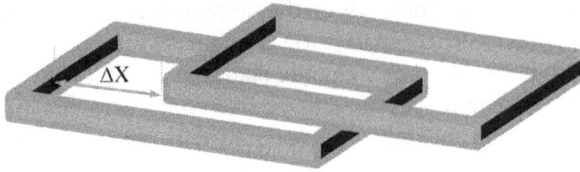

Figure 6.2 Schematic of studied variation of horizontal misalignment.

this form of misalignment (Figure 6.2). When the secondary coil shifts along the x-axis with the plane parallel to the primary coil plane at a specified height, horizontal misalignment occurs.

6.4 ANALYSIS OF MAGNETIC CIRCUIT FOR DD COIL STRUCTURE

The DD coil of WPT has a high coupling coefficient and a greater tolerance for lateral offset. Figure 6.3 depicts the design of the DD coils, and Figure 6.4 illustrates the DD coil with ferrite core. The axis of the DD coil is used to simplify the analysis as can be seen in Figure 6.5 shows the magnetic coupling distribution Circuit. The DD coil on the transmitter contains two coils, one for each channel and considered as having independent sources known as electromotive forces P_1 and P_2.

The magnetomotive forces of the two transmitting coils are labelled P_1 and P_2. The self-reluctances of the coils are labelled R_{s1}, R_{s2}, & R_{s3}. The mutual reluctance of the coils are R_{m1}, R_{m2} and α_{s1}, α_{s2}, and α_{s3} represent the flux produced by the self-linking of the magnetic field are the fluxes produced by the mutual linking of the magnetic field. Using the magnetic circuit model's expression, the mathematical expression of magnetic fluxes of each type can be obtained.

$$\alpha_{m1} = \frac{P_1 + P_2}{R_{m1}} \tag{6.1}$$

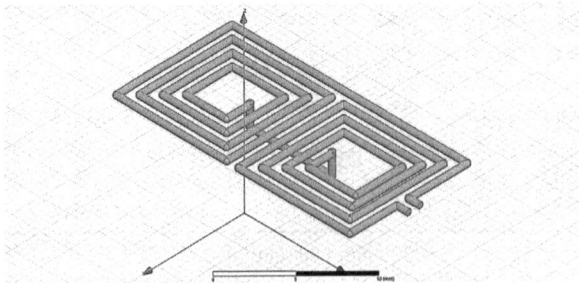

Figure 6.3 Simple presentation of DD coil structure.

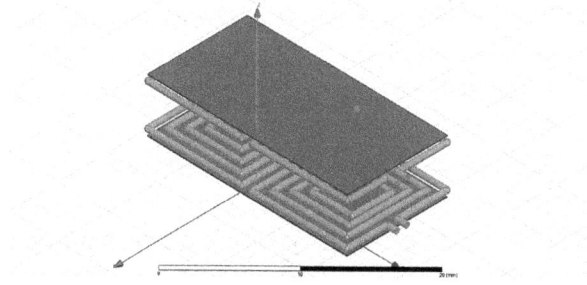

Figure 6.4 A simple structure of DD coil with ferrite.

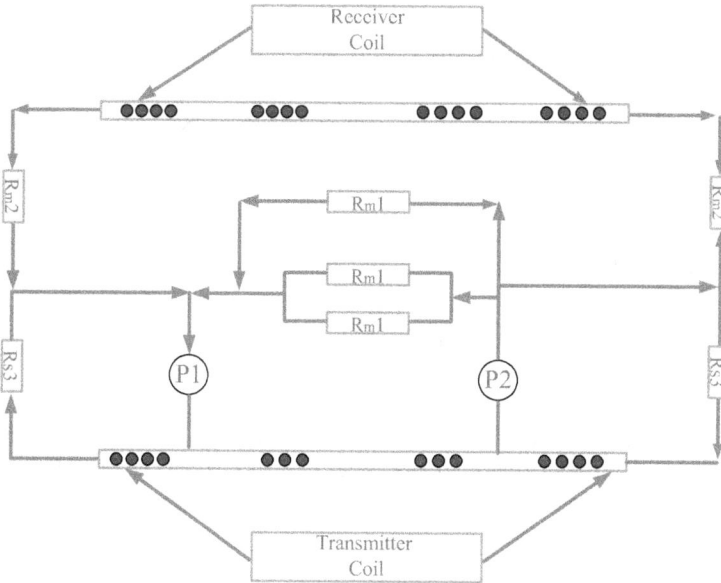

Figure 6.5 Magnetic circuit of DD coil structure.

$$\alpha_{m2} = \frac{P_1 + P_2}{2R_{m2}} \tag{6.2}$$

$$\alpha_{s1} = \frac{P_1}{R_{s3}} \tag{6.3}$$

$$\alpha_{s2} = \frac{P_1 + P_2}{R_{s1} + R_{s2}} \tag{6.4}$$

$$\alpha_{s3} = \frac{P_2}{R_{s3}} \tag{6.5}$$

Coupling coefficient (k)

$$k = \frac{\alpha_{m1} + \alpha_{m2}}{\alpha_{m1} + \alpha_{m2} + \alpha_{s1} + \alpha_{s2} + \alpha_{s3}} \qquad (6.6)$$

$$k = \frac{\dfrac{P_1 + P_2}{R_{m1}} + \dfrac{P_1 + P_2}{2R_{m1}}}{\dfrac{P_1}{R_{s3}} + \dfrac{P_1 + P_2}{R_{s1} + R_{s2}} + \dfrac{P_2}{R_{s3}} + \dfrac{P_1 + P_2}{R_{m1}} + \dfrac{P_1 + P_2}{2R_{m2}}}$$

As well as the symmetrical structure and relationship between the two corresponding Coil D, $P_1 = P_2$.

The equation for the coupling coefficient can be written as follows:

$$k = \frac{1}{\dfrac{\psi_1 + \psi_{s1}}{\psi_{m1} + 1/2\psi_{m1}} + 1} \qquad (6.7)$$

Where $\psi = 1/R$

$$\psi = \frac{1}{R_{s1} + R_{s2}} \qquad (6.8)$$

We can conclude from Equation (6.7) that the coupling coefficient is dependent on the placement of the core material, since the coupling coefficient is improved by simultaneously reducing $(\psi_{s3} + \psi)$ and increasing $(\psi m1 + 0.5\ \psi m2)$.

6.5 RESULT AND SIMULATION

A 3-D FEA simulation has been done in Ansys Maxwell® 2015 to validate the proposed model. The model is based on the parameters given in Table 6.1. In the simulation, the size of the magnetic shield has been taken comparatively larger than the coil dimension, because of the assumption of the infinite size of the magnetic shield in the model. The primary coil is made of copper and the shield is made of ferrite.

The proposed model is used to calculate the mutual inductance and the coupling coefficient for dissimilar misalignment positions. The comparison of variation in mutual inductance, coupling coefficient and misalignment is achieved from FEA shown in Table 6.2. Figure 6.6 shows the graphical representation of coupling coefficient variation with respect to the misalignment in the x-axis. As the secondary coil moves along the x-axis, the mutual

Table 6.1 Parameters of coil system

Parameter	Values
Air gap distance	0.5 mm
Tx coil current	10 A
Rx coil current	10 A
Number of coil turn	4
Diameter of coil	0.5 mm
Size of coil	20*10 mm

Table 6.2 Comparative results for different misalignment position

X misalignment	Coupling coefficient (k)	Mutual inductance	Magnetic flux (Wb)
00	0.2374	72.699280	0.000004
2	0.1669	52.111430	0.000004
4	0.07612	23.894770	0.000003
6	0.001045	326.83460 (PH)	0.000003
8	0.040534	12.565370	0.000003
10	0.046119	14.177230	0.000003

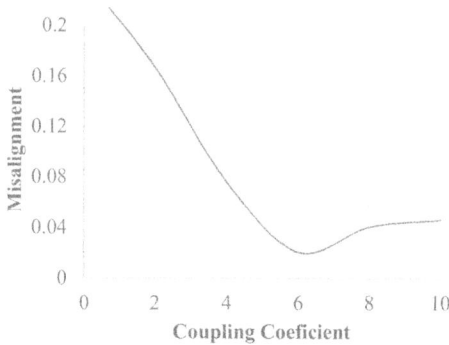

Figure 6.6 Coupling coefficient variation for misalignment.

inductance decreases. This decrease in mutual inductance is because of the decrease in the magnitude of the magnetic field density with respect to the x-axis direction. Figure 6.7 shows the different misalignment position in the x-axis direction.

Figure 6.8 illustrates the variation in the mutual inductance caused by vertical misalignment. As a result of the decrease and then the increase in the overlapping area between the coils, the mutual inductance reduces for 0 to 10 mm of secondary coil and then rises for 10 to 0 mm misalignment.

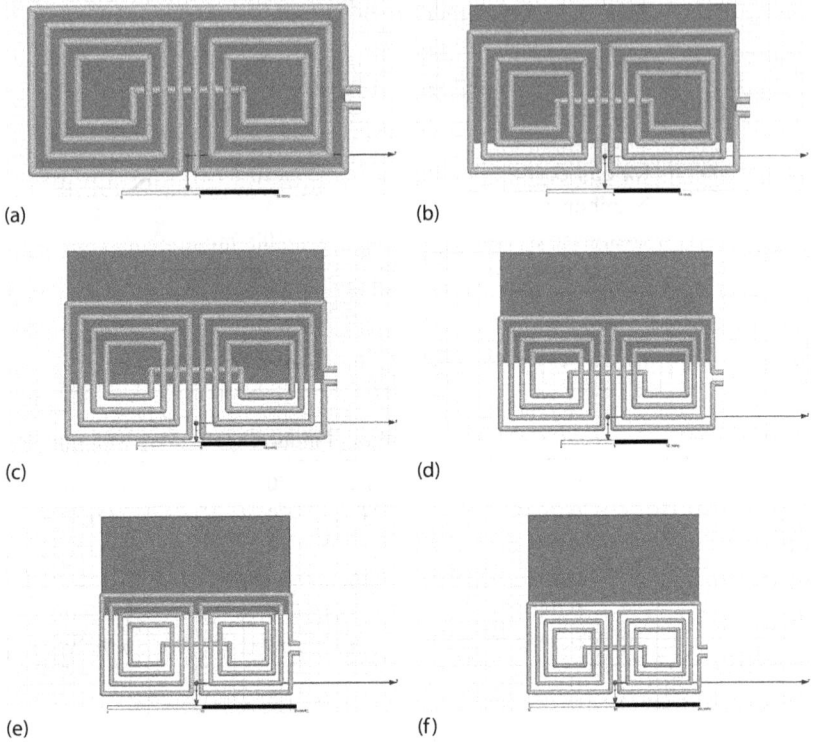

(a)

(b)

(c)

(d)

(e)

(f)

Figure 6.7 Misalignment position.

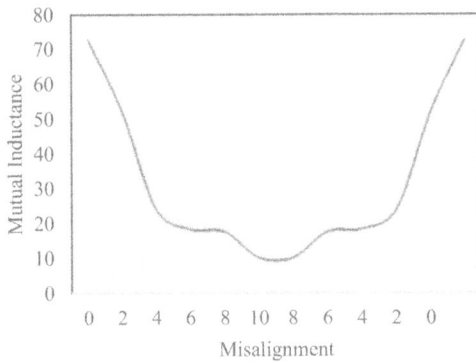

Figure 6.8 Mutual inductance variation in x-direction.

6.6 CONCLUSION

In this chapter, an analytical model is presented to compute the mutual inductance and coupling coefficient between the coils for different misalignment for a coil system having a magnetic shield below the primary and above the secondary coil. Further, a simulation setup with different misalignment is developed in the Ansys Maxwell® 2015 to validate the efficacy of the model. Graphical representations of both the coupling coefficient and mutual inductance with different misalignment is proposed, meaning that the proposed model can be used in the initial design of such a dynamic wireless power transfer system.

ACKNOWLEDGEMENTS

This research was funded by the collaborative research grant scheme (CRGS) project, Hardware-In-the-Loop (HIL) Lab, Department of Electrical Engineering, Aligarh Muslim University, India having project numbers CRGS/MOHD TARIQ/01 and CRGS/MOHD TARIQ/02.

The authors also acknowledge the technical support provided by the Hardware-In-the-Loop (HIL) Lab, Department of Electrical Engineering, Aligarh Muslim University, India.

REFERENCES

[1] A. Ahmad, M. S. Alam, Y. Rafat, S. M. Shariff, I. S. Al-Saidan, and R. C. Chabaan, "Foreign object Debris detection and automatic elimination for autonomous electric vehicles wireless charging application," *SAE Int. J. Electrified Veh.*, vol. 9, no. 2, pp. 93–110, 2020, doi:10.4271/14-09-02-0006

[2] M. Budhia, G. Covic, and J. Boys, "A new IPT magnetic coupler for electric vehicle charging systems," *IECON 2010 - 36th Annual Conference on IEEE Industrial Electronics Society*, 2010, pp. 2487–2492, doi: 10.1109/IECON.2010.5675350

[3] Wajahat Khan, Aqueel Ahmad, Furkan Ahmad, and M. Saad Alam, "A comprehensive review of fast charging infrastructure for electric vehicles," doi:10.1080/23080477.2018.1437323

[4] A. Ahmad, M. S. Alam, and R. Chabaan, "A comprehensive review of wireless charging technologies for electric vehicles," *IEEE Trans. Transp. Electrif.*, vol. 4, no. 1, pp. 38–63, 2017, doi:10.1109/TTE.2017.2771619

[5] T. Abdelouahed and Z. S. Ahmed, "Modeling and transient simulation of unified power flow controllers (UPFC) in power system," *2015 4th Int. Conf. Electr. Eng. ICEE 2015*, no. March 2021, 2016, doi:10.1109/INTEE.2015.7416851

[6] A. A. S. Mohamed, A. A. Shaier, H. Metwally, and S. I. Selem, "Interoperability of the universal WPT3 transmitter with different receivers for electric vehicle inductive charger," *eTransportation*, vol. 6, no. September 2020, p. 100084, 2021, doi:10.1016/j.etran.2020.100084

[7] Bilal Alam, Maaz Nusrat, Zeeshan Sarwer, and Mohammad Zaid, "A general review of the recently proposed asymmetrical multilevel inverter topologies," *Innovations Cyber Phys. Syst. Lect. Notes Electr. Eng.*, doi:10.1007/978-981-16-4149-7_61

[8] V. Shevchenko, O. Husev, R. Strzelecki, B. Pakhaliuk, N. Poliakov, and N. Strzelecka, "Compensation topologies in IPT systems: Standards, requirements, classification, analysis, comparison and application," *IEEE Access*, vol. 7, pp. 120559–120580, 2020, doi:10.1109/ACCESS.2019.2937891

[9] M. Brenna, F. Foiadelli, C. Leone, and M. Longo, "Electric vehicles charging technology review and optimal size estimation," *J. Electr. Eng. Technol.*, vol. 15, no. 6, pp. 2539–2552, 2020, doi:10.1007/s42835-020-00547-x

[10] M. L. NarayanaGadupudi et al., "Recent advances of STATCOM in power transmission lines – A review," *Turkish J. Comput. Math. Educ.*, vol. 12, no. 3, pp. 4621–4626, 2021, doi:10.17762/turcomat.v12i3.1864

[11] G. Wang and J. Sun, "Improved magnetic coupling resonance wireless power transfer system," *Chinese Control Conf. CCC*, vol. 2020-July, pp. 5317–5321, 2020, doi:10.23919/CCC50068.2020.9189653

[12] Q. Chen, S. C. Wong, C. K. Tse, and X. Ruan, "Analysis, design, and control of a transcutaneous power regulator for artificial hearts," *IEEE Trans. Biomed. Circuits Syst.*, vol. 3, no. 1, pp. 23–31, 2009, doi:10.1109/TBCAS.2008.2006492

[13] A. Kamineni, G. A. Covic, and J. T. Boys, "Analysis of coplanar intermediate coil structures in inductive power transfer systems," *IEEE Trans. Power Electron.*, vol. 30, no. 11, pp. 6141–6154, 2015, doi:10.1109/TPEL.2014.2378733

An automated system for the rapid classification of harmonic loads and power system faults

A. Pullabhatla Srikanth and Chiranjib Koley
NIT, Durgapur, India

CONTENTS

7.1 INTRODUCTION: BACKGROUND AND DRIVING FORCES

Over the past decade, extensive research has been carried out into power system quality analysis, identification, localisation, and remedial actions. The researchers have worked consistently on applying signal processing and artificial intelligence techniques to identify power system abnormalities. Such research is a two-step process wherein the first step is to extract the critical parameters for identification. The second is the type of techniques used for identification, classification or localisation. The extraction of parameters is crucial as it decides the computational burden for successfully identifying and classifying PQ events.

Both parametric and non-parametric methods were proposed for extracting the features of any given signal. Parametric methods include ESPIRIT, MUSIC, ARMA, Auto-Regressive, and Prony analysis, whereas non-parametric methods include analysis based on Signal Processing methods. In the past power system disturbances have been identified and analysed effectively using the parametric methods; however, the applications were limited. It is observed that most of the parametric methods are appropriate for post-fault or post-disturbance analysis and for determining what type of disturbance has occurred. The methodology lacks online monitoring with high accuracy [1–6].

DOI: 10.1201/9781003311195-7

Non-parametric approaches were also adopted by many researchers for the identification of power quality events. Techniques such as Wavelet Transforms, Stockwell Transforms, and Gabor-Wigner transform, and modified versions of these techniques, were used to identify the type of power quality event. s Z. Moravej et al. have proposed a wavelet transform-based method with a support vector machine for recognising and classifying power quality events like Sag, Swell, Transients, etc. However, the authors have identified the signal based on standard mathematical equivalents of these power quality events. Furthermore, the authors have not discussed the power system faults and the origin of all the disturbances considered [7]. In another study, Haibo He et al. used an Energy Difference Multi Resolution Analysis (EDMRA) method to identify the power quality events. The authors have compared various types of wavelets and found that EDMRA is more robust and scalable than the original wavelet transform. However, the authors have identified only low frequency, high frequency, sag distortion and swell distortion in the paper [8].

Similarly, Mojtaba Kordestani et al. have also proposed a different version of wavelets to identify the type of PQ event but limited only to harmonics [9]. Wavelets and Artificial Neural Networks (ANN) were proposed to determine the power system faults by Paul Malla et al. However, the results obtained were on a single machine system, and three-phase faults were not identified [10]. The Stockwell transform (ST) was also used to determine the type of power quality event, including power system faults by Maddikara Jaya Bharata Reddy et al. A LabVIEW-based model was presented in the paper. However, the problem of localisation of the type of PQ event was not discussed [11].

Moravej et al. have proposed ST with LMT, Hasheminejad et al. have proposed ST with a Markov model, and Milan Biswal et al. have proposed ST with a decision tree [12–14]. However, they have not identified the faults, and the results were presented on mathematical equations or the 2-Bus system. Ali Enshaee et al. have also proposed an ST-based methodology for identifying PQ events by excluding power system faults and localising PQ event sources. The results include a wide variety of PQ events with the help of ST and their extracted parameters. However, the number of parameters defined for identification are higher [15].

The above discussion of the background literature clearly shows that a system with both the identification and localisation of PQ events, including faults, is not reported for an interconnected power system. However, integrated PQ monitoring systems were proposed by some of the researchers. An integrated PQ monitoring system was introduced by Music et al., in which the authors implemented the PQMS on a substation installed in Bosnia and Herzegovina. However, the application was limited to current harmonics only. The authors have used a standard Integrated Power Quality Monitoring system (IPQMS) to identify PQ events and have not discussed the computational burden on the proposed system [16]. Cheng-I Chen et al. have introduced a new IPQMS based on a sliding window method and have identified all types of harmonics for a given microgrid. However, the case of power system faults was not considered in the paper. In addition, there is no discussion about the source of harmonics and localisation of the PQ events [17].

In totality, it is clear that there is still a need to develop an integrated system that can identify, classify, and localise the type of PQ event and its source. The system should be adaptable to any interconnected power system, learn changes in the system with a minimum computational burden, and maintain high levels of accuracy. The development of such a system forms the objective of this article and accounts for the central part in the problem formulation. To design such a system, the authors have proposed a time-frequency-based methodology and a Support Vector Machine for the easy identification and localisation of PQ events and their sources. The methodology used for parameter extraction is explained in Section 7.2. The extracted parameters are defined in Section 7.3. The design of the identification and localisation system using SVM is presented in Section 7.4. Next, Section 7.5 is about the results obtained and the discussions. Section 7.6 is included for comparisons and conclusions.

7.2 SIMULATION MODEL

An IEEE-9 Bus system has been used to obtain the waveforms. It was ensured that all the parameters were considered as defined by the IEEE working group during the simulation. Both types of faults, i.e. symmetrical and unsymmetrical faults, have been simulated. The fault inception length and location has been changed to obtain waveforms at all the three recording stations, i.e. at Bus no. 4, 7 & 9 in the IEEE-9 Bus system. The faults were simulated on the transmission line between Bus 4 & 5, Bus 8 & 9 and Bus 7 & 8. In total, three combinations have been considered for the identification and location of faults.

Further, nonlinear industrial harmonics loads have been modelled and fed into the power system. Arc Furnace and Electric traction drives have been modelled using the data as defined by Vinayaka et al. and the ABB Guide to Harmonics in AC drives [18, 19]. These loads are found very commonly in any 11 kV or 33 kV industrial systems. The localisation of harmonics sources such as Arc Furnace and Traction loads is critical as they cause voltage flickers and unbalanced currents with harmonics into the systems.

In addition to the above, the detection of AC drive loads is also considered, as industries such as cement and steel use huge VFDs/AC drives with negligence towards PQ filtering. These cause harmonic injection along with sags in the power supply. Other than these nonlinear loads, events like Swell+Harmonics and only Harmonics are also considered per the predefined models, thus considering all types of harmonics present in the power system. Figure 7.1 shows an IEEE standard nine-Bus system, where all the quantities are represented per unit (PU). EMTP has been used to simulate

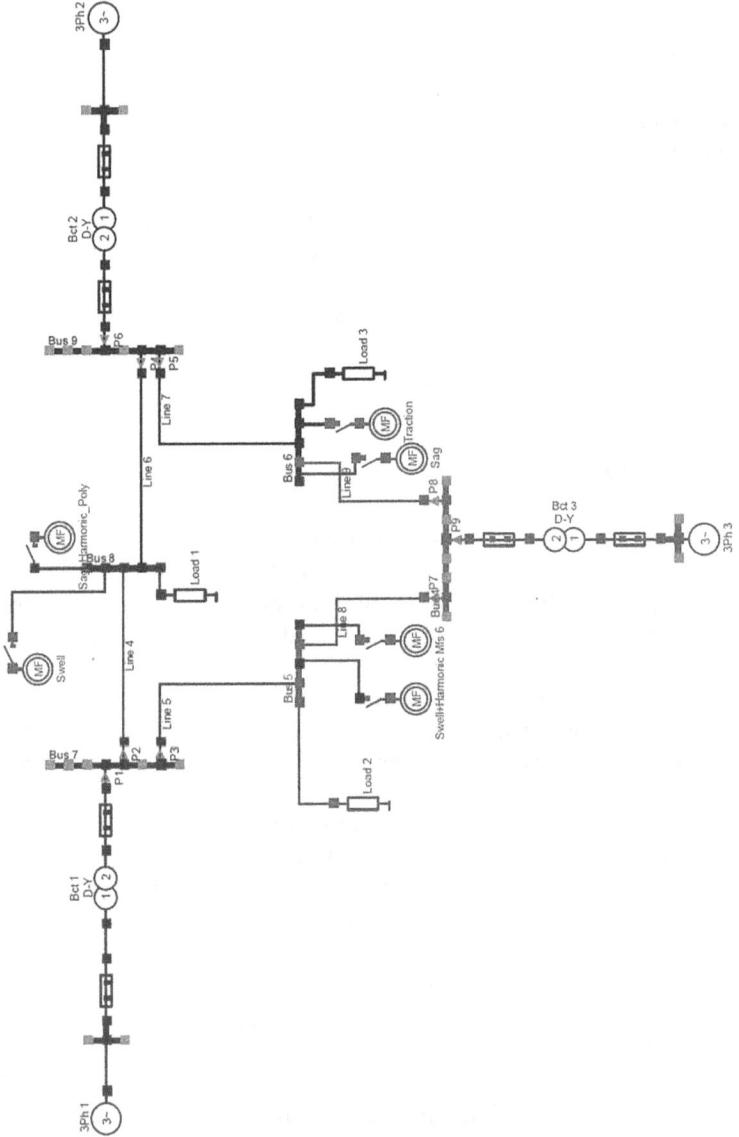

Figure 7.1 EMTP model of IEEE 9-Bus along with various case studies operating singly at any instant.

the model [20]. Firstly, power system faults are simulated, and then the modelled harmonics sources are fed at each Bus for the identification process. All disturbances are introduced from t = 0.06s to 0.12s. The recorded currents in the measurement units are used to extract the time-frequency properties and further analysis. The currents are measured per unit, and the same is fed to the algorithm identification.

7.3 METHODOLOGY AND PARAMETER EXTRACTION

7.3.1 Discrete Stockwell-Transform (DST) methodology

Since Discrete S-Transform (DST) represents the local spectra, it can be obtained by the shift operation on the Fourier spectrum and is expressed as Equation (7.1), which has been well defined and formulated by many researchers in the past.

$$ S\left[kT, \frac{n}{NT} \right] = \sum_{k=0}^{N-1} H\left[\frac{m+n}{NT} \right] \cdot e^{\frac{-2\pi^2 m^2}{n^2}} e^{\frac{i2\pi mk}{N}} \tag{7.1} $$

where $k, n, m = 0 .... N - 1$, T = sampling time, and N = number of samples.

It is known that S-Transform (ST) produces the localisation of both phase and amplitude spectrum. The Gaussian window defined in ST varies according to both the time and frequency of the input signal. Because of such a definition, ST maintains good time and frequency resolution. The power system disturbances being non-stationary, the Stockwell transform can be applied effectively. The next section discusses the extraction of discriminatory features of the ST matrix coefficients obtained from Equation (7.1) for the identification of the source of PQ disturbance.

7.3.2 Parameter extraction

Since the ST response is exclusive for each type of fault, the ST matrix values, as shown in Equation (7.1), have been used for the parameter extraction. The properties of the ST matrix are shown in Figure 7.2. The ST matrix contains the responses at various times and frequencies, known as locals and voices. A local is obtained when the time is constant with variable frequency, and a voice is a variation in response to the change in time at a constant frequency. Both these properties of a time-frequency matrix provide much helpful information. A local gives the information about variations in amplitudes at any instant, and a voice provides the information of variation in amplitudes at a particular frequency. The same is demonstrated with a sample ST plot of power system signal disturbances in Figure 7.2(a) and (b). For a pure sinusoidal signal, the maximum values of the locals at any time are the same, and the voices are the same beyond the fundamental frequency, as there are no additional frequency components/magnitude disturbances.

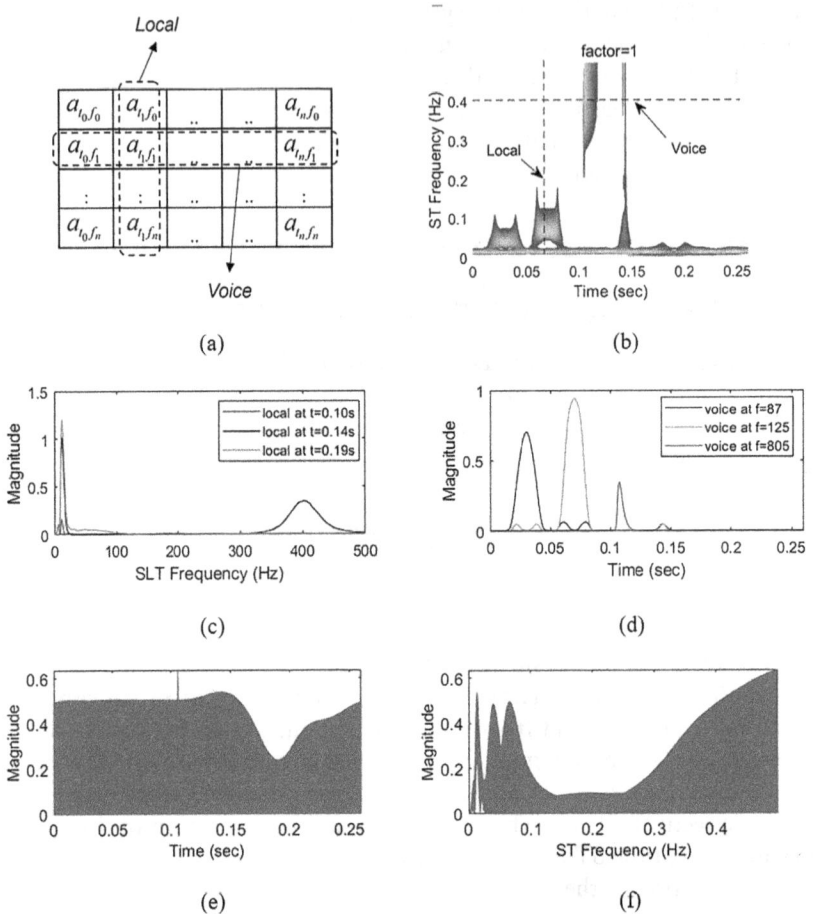

Figure 7.2 Elements of a ST matrix with the definition of voice and local. (a) ST matrix formation (b) representation of the ST response plot as defined in Equation (1) (c) locals at various intervals (d) voices at various frequencies (e) shaded area under the curve of maximum values of all locals (ACM) (f) shaded area under the curve of maximum values of all voices (ARM).

However, for a power system signal with disturbances, the maximum values of locals and voices change according to the frequencies and magnitudes introduced by the disturbances. Hence, these maximum values of each local and voice naturally contain essential information about power system disturbances. The plot of highest magnitudes of all locals vs. frequency, i.e. the maximum values of columns of the ST matrix in Figure 7.2(b) vs. ST frequencies (CM) and the plot of highest magnitude of all voices in Figure 7.2(b) vs. time, i.e. the maximum values of all rows of the ST matrix vs. ST time (RM), provide the information of the variations of an input signal.

The locals and voices at various time and frequency occurrences of the ST matrix in 2(b) are shown in Figures 7.2(c) and (d). These locals and voices indicate the presence of other frequency components and magnitude variations. Figures 7.2(e) and (f) are the area plot of the maximum values of each local and voice of the ST matrix. There are variations in the magnitudes in the RM plot, and the reoccurrence of the peak is due to the events like the voltage regain after an interruption at 0.2 s in Figure 7.2(e). At the same time, the area under the maximum values of each voice indicates the presence of multiple peaks, i.e., high-frequency components. The area under these CM and RM plots has been used to identify the type and location of the power quality events. Below mentioned are the parameters defined for the identification of the type and location of power system faults.

$$ACM = Area[max. \text{ values of Locals of ST matrix}] \qquad (7.2)$$

$$ARM = Area[max. \text{ values of Voices of ST matrix}] \qquad (7.3)$$

The area obtained using the MATLAB® function *trapz* with the y-axis is the ST Magnitude and the x-axis is either ST Time or Frequency. The features, i.e. the values of ARM and ACM, are obtained for different types of faults at a different distance of occurrence, i.e. 20%, 40%, 60% & 80% of the T/m line length and recorded at Bus 4, Bus 7 and Bus 9 in the IEEE-9 Bus system. In addition, cases like Sag+Harmonics, Swell+Harmonics, Arc Furnace Loads, 12-Pulse Rectifier and Traction loads are considered on the same buses for automatic identification and localisation. The overall identification process is shown in Figure 7.3.

7.4 RESULTS AND DISCUSSIONS

This section presents the obtained results with the methodology explained in the previous sections. The simulated model, as shown in Figure 7.1, has been used for obtaining the results. The PQ disturbances which were considered for identification, i.e. symmetrical faults (3Φ), asymmetrical faults (2Φ & 1Φ), arc furnace load, electric traction load, industrial VFD fed from a 12-Pulse Double wound XFMR causing Sag+Harmonics, Swell+Harmonics and only Harmonics are introduced one at a time. A robust data set of power system faults have been recorded at 20 km, 40 km, 60 km and 80 km from Bus 4, Bus 7 and Bus 9 as the location of the faults may vary. The EMTP model of the IEEE 9 Bus system is fed with each case singly. Three cycles of the recorded current in p.u. has been fed to the ST algorithm, and the parameters as defined in Section 3.2 are obtained. The current waveforms recorded at Bus 4 and the ST plot, ACM and ARM plots described previously are shown in Figures 7.4 and 7.5. Electric Traction load at Bus 5

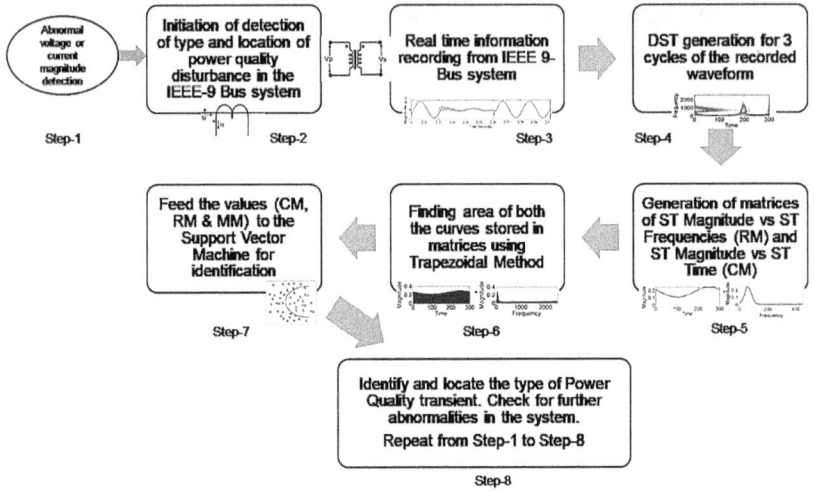

Figure 7.3 Graphical representation of the overall process of location and identification of power system faults.

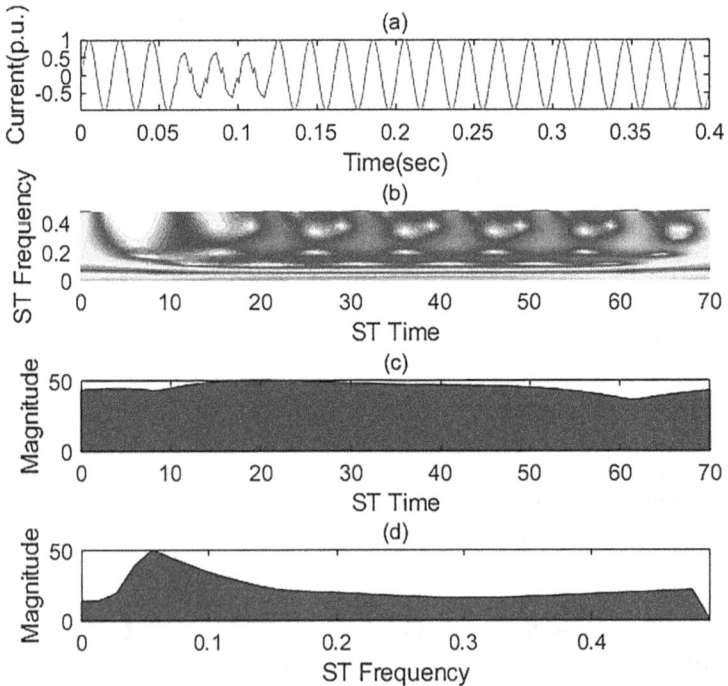

Figure 7.4 (a) Current waveforms recorded at Bus 4 along with the (b) ST plot, (c) ACM Plot and (d) ARM plots for Electric Traction load at Bus 5.

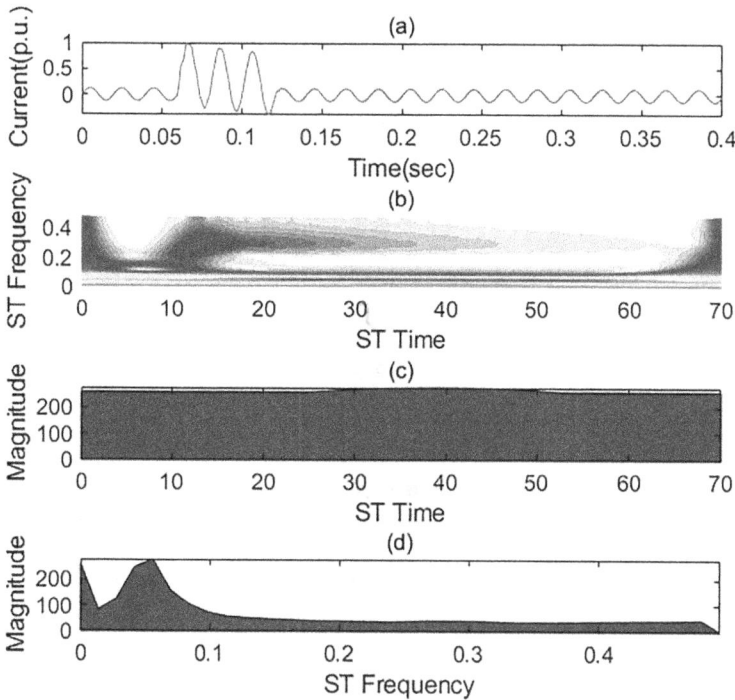

Figure 7.5 (a) Current waveforms recorded at Bus 9 along with the (b) ST plot, (c) ACM Plot and (d) ARM plots for AB fault at a distance of 60 km from Bus 6.

and AB fault at a distance of 60 Km from Bus 6 are shown in Figures 7.4 and 7.5, respectively.

It is apparent that the variation in the magnitude for the harmonic loads and power system faults is different. However, it was found that the waveforms of the power system faults group are almost similar for various types of faults. The waveforms of the harmonic loads' group are also similar though the nature of harmonic loads is different. Hence, highly discriminative features are required for 100% accurate identification of the type of PQ disturbance and its origin. A complete data set has been obtained for both the defined features, i.e. ACM and ARM, using the proposed methodology. These values are presented in Table 7.1. It was observed that the area of ARM is at a higher range in comparison to ACM. The x-axis for calculating ARM is the frequency that varies from 50 to 500 Hz, i.e. 0.05 to 0.5 Hz in the ST domain. In contrast, the x-axis for ACM is time in seconds, i.e. $t = 0.06$ to 0.12s, shown as 0 to 70 samples in 7.4(b) and 7.5(b). However, both ACM and ARM values are normalised to unity as the proposed system is for universal application, i.e. it shall be suitable for any base power supply frequency and any level of voltage.

Table 7.1 Area under the envelope of maximum value of each row
of ST matrix w.r.t frequency for different cases of PQ
disturbances originated at Bus 5 and recorded at Bus 4

Type of PQ disturbance	*Normalised area of envelope of maximum values of rows (ARM) for different type of faults*	*Normalised area of envelope of maximum values of columns (ACM) for different type of faults*
Harmonics	0.32586	0.19109
Arc Furnace Load	0.25271	0.12300
Electric Traction Load	0.24087	0.12329
12P Double Wound XFMR	0.26139	0.12243
AB_Fault at 20 Km	0.78690	0.67249
AB_Fault at 40 Km	0.81077	0.71236
AB_Fault at 60 Km	0.83002	0.74794
AB_Fault at 80 Km	0.88878	0.80528
ABC Fault at 20 Km	0.90003	0.79823
ABC Fault at 40 Km	0.93670	0.85835
ABC Fault at 60 Km	0.97508	0.92494
ABC Fault at 80 Km	0.99977	1.00000
ABG Fault at 20 Km	0.87089	0.69935
ABG Fault at 40 Km	0.89916	0.73975
ABG Fault at 60 Km	0.92386	0.77725
ABG Fault at 80 Km	1.00000	0.83684
AG Fault at 20 Km	0.70271	0.45023
AG Fault at 40 Km	0.72323	0.45759
AG Fault at 60 Km	0.74037	0.46887
AG Fault at 80 Km	0.75983	0.48369
Swell+Harmonics	0.31383	0.23641

Table 7.1 indicates that the extracted feature, i.e. ARM, is dynamic and behaves in a manner similar to the power system transient variation. The ARM values are high for the faults near the substation, i.e. the Bus, and are low for the faults occurring further away. The pattern is the same for all types of faults. The values of ACM also have the same pattern. In contrast, the values of harmonics loads are less than the power system faults, making them easier to identify.

Similarly, the values of ARM and ACM for Bus 7 and Bus 9 are also obtained to feed the SVM. As a whole, the values of ARM and ACM are unique for all types of PQ disturbances at each Bus. It is evident that the extracted features from the proposed area under the curve approach fit the identification of the type of faults.

The next step in identifying PQ disturbance is to feed the obtained values into an identification algorithm. In the present work, a Support Vector

Machine (SVM) is used for multiclass identification problems. Both linear and nonlinear SVMs have been explored to find the best fit kernel for identifying power quality disturbances. All types of kernels, viz. Linear, Quadratic, Cubic, Fine & Coarse Gaussian, are used to check the efficiency of the values extracted. In order to test the algorithm 100 sets of data have been used, whereas three (20) sets were used for validation. The training of the SVM is done through the MATLAB R2018a version [21]. As mentioned earlier, the waveform data is obtained from EMTP as the simulation results are very accurate and near to real-time in nature. The SVM is chosen to identify because it is easy to compute, faster prediction is possible, and it deals efficiently with higher dimensionality. As more cases are considered for identification, the SVM proves an appropriate tool.

Figure 7.6 shows the SVM classifier window indicating eight classes represented on the ARM vs. ACM plane. For better accuracy and training, the cross-validation method has been used. In the present case, 50-fold cross-validation is considered. It is visible from the SVM classification plot that power system faults and harmonics are discriminated against and classified with the respective source. The type of power system faults, i.e. 3-Phase, 2-Phase and 1-Phase faults, are identified. In addition, the cause of flicker, unbalance and Sag+Harmonics, i.e. Arc Furnace and Electric Traction, VFD, are easily identified. All three cases of Harmonics, i.e. Sag+Harmonics, Harmonics and Swell+Harmonics, flicker and imbalance, are successfully identified without overlap or confusion. It indicates that the parameters extracted for the subject work are well defined and discriminative. The performance results of the SVM classification with different kernels are shown in Table 7.2. The classification accuracy of the proposed ST+SVM for the IEEE-9 Bus system at Bus 4 is 100% with 8 types of PQ disturbances.

Similarly, the classification of PQ recorded at Bus 9 originated at Bus 6, including faults between Bus 6 and Bus 9, has also been performed using the SVM. The classification accuracy, in this case, is also found to be 100%. The performance of the proposed algorithm implemented at Bus 9 is shown in Table 7.3. It is observed that Fine Gaussian kernel SVM is also more efficient in this case. The results obtained for the algorithm implemented at Bus 7 are shown in Table 7.4. The SVM classification has also identified all the PQ disturbances accurately in this case. The Fine Gauss SVM has shown 100% accuracy. Evidently, there is no confusion in identifying the type of PQ disturbance.

For all three cases, it is observed that the proposed methodology of ST+SVM is very efficient, fast and highly accurate. The trained algorithm takes only <0.4 s to detect and locate the type of PQ disturbance. This makes the proposed algorithm suitable for implementation in any power quality monitoring equipment in real time. The Fine Gauss-based SVM has provided 100% accuracy in all the cases with the fastest training time, making it the most suitable kernel for the proposed methodology. Further, a comparison with the previously proposed methodologies is also made and

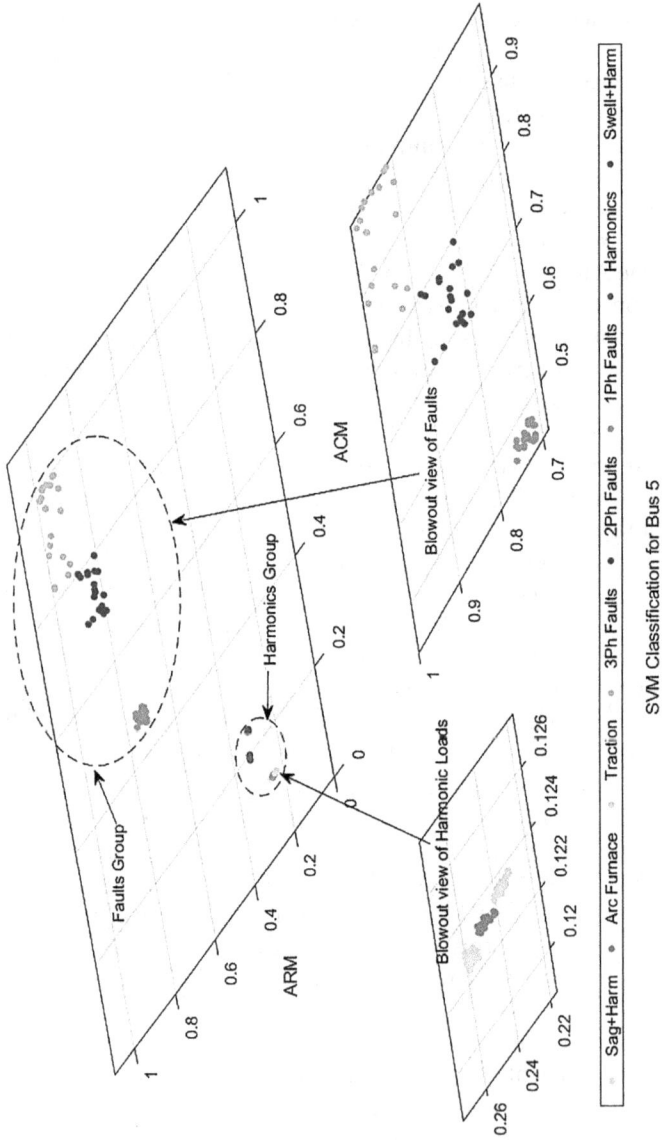

Figure 7.6 SVM classification of various PQ disturbances along with the source for Bus 5 with Blowout view of Harmonics and fault events.

Table 7.2 Comparison of performance of SVM classification algorithms at Bus 4

S. no	Type of SVM	Training time (in sec)	Accuracy (%)
I	Linear SVM	7.99	98.9
2	Quadratic SVM	7.70	99.4
3	Cubic SVM	8.22	99.4
4	Fine Gaussian SVM	7.65	100
5	Medium Gaussian SVM	7.93	100
6	Coarse Gaussian SVM	7.80	100

Table 7.3 Comparison of performance of SVM classification algorithms at Bus 7

S. no	Type of SVM	Training time (in sec)	Accuracy (%)
I	Linear SVM	7.64	96.1
2	Quadratic SVM	7.67	98.9
3	Cubic SVM	7.50	98.9
4	Fine Gaussian SVM	7.29	100
5	Medium Gaussian SVM	7.25	99.4
6	Coarse Gaussian SVM	7.23	93.9

Table 7.4 Comparison of performance of SVM classification algorithms at Bus 9

S. no	Type of SVM	Training time (in sec)	Accuracy (%)
I	Linear SVM	7.64	96.1
2	Quadratic SVM	7.67	98.9
3	Cubic SVM	7.50	98.9
4	Fine Gaussian SVM	7.29	100
5	Medium Gaussian SVM	7.25	99.4
6	Coarse Gaussian SVM	7.23	93.9

presented in Table 7.5. For comparison, articles from the last 10 years reporting the application of ST+Artificial intelligence techniques to identify PQ disturbances are considered.

These articles are referred to at [12–15]. The reason for considering these articles is that the accuracy levels reported are very high, and the application is similar. The proposed methodology in this essay is different in that a realistic power system scenario is showcased. The effects of generator impedance, transmission line impedance, load impedance, harmonic load models and measurement noise are included while recording the current waveforms. This has not been reported in the literature to date, as seen in Table 7.5,

Table 7.5 Comparison with the results reported in the literature in the last 10 years using ST and artificial intelligence technique

Ref. no.	Methodology used	Mathematical equivalent wave forms used (Yes/No)	Interconnected power system used (Yes/No)	Noise considered (Yes/No)	Number of parameters extracted (Yes/No)	Faults Identified along with PQD (Yes/No)	Harmonic loads Identified (Yes/No)	Location of PQ disturbances identified (Yes/No)	Accuracy reported (%)
[12]	ST+LMT	Yes	No	Yes	04	No	No	No	99.11
[13]	ST+Markov model	No	Yes 2-Area system	Yes	03	No	Yes	No	98.14
[14]	ST+Decision tree	Yes	No	Yes	20	No	No	No	98.80
[15]	ST+Decision tree	Yes	No	Yes	07	No	No	No	99.39
This article	ST+SVM	No	Yes IEEE-9 Bus	Yes	02	Yes	Yes	Yes	100

where only one article has considered a 2-Bus interconnected power system. The proposed methodology of extracting parameters in Section 3.3 is also novel, with only two defined parameters. With these factors, the proposed methodology of ST+SVM can produce 100% accuracy.

7.5 CONCLUSIONS

This chapter presents a novel approach to identifying and locating the type of PQ disturbance by implementing the algorithm at one of the measurement points. An IEEE-9 Bus system, simulated in EMTP, is used to record the data at three different buses connected to the generating stations. Eight types of PQ disturbances viz. 3-Phase, 2-Phase, 1-Phase faults, Harmonics, Swell+Harmonics, Sag+Harmonics due to VFD, current flickers due to Arc Furnace, imbalance due to Electric Traction are designed and simulated to obtain the current measurements. These disturbances are introduced in a farther location, which is 100 Km from the recording bus. A total of three buses are chosen for implementing the algorithm, i.e. Bus 4, Bus 7 and Bus 9, wherein the disturbances were created at Bus 5, Bus 8 and Bus 6, respectively.

The proposed algorithm of ST+SVM has been implemented at the recording bus, thus creating a scenario of PQ monitoring in a real power system. All impedances affecting the system current, i.e. the impedances of the supply, transformer, transmission line, load and harmonic load, are accommodated to obtain a realistic waveform. Later, the recorded data is fed to the ST algorithm, wherein the defined features based on the area under the envelope method are extracted. These features are fed to the SVM tool for identification. It was observed that the implementation of the SVM had produced 100% accuracy in the identification and location of the PQ disturbance. Fine Gauss kernel-based SVM has provided 100% accuracy in all three cases. The algorithm has identified the location of the disturbance, along with the nature of the PQ disturbance with only two defined parameters, which reduces the computational burden. It has separated the Harmonics and Fault events, thus serving as a robust platform for classifying high- and low-magnitude transients like faults and harmonics. The proposed algorithm can be implemented in a real-time PQ monitoring device for larger bus systems, as the defined features and the number of PQ events remain the same.

REFERENCES

[1] Gajjar, Gopal, and S. A. Soman. "Power System Oscillation Modes Identifications from Wide Area Frequency Measurement System." *2012 IEEE International Conference on Power System Technology (POWERCON)*, 2012, pp. 1–6. IEEE Xplore.

[2] Shang, Li-qun, and Jian-dang Lv. "A New Approach for Identification of the Fault Type on Transmission Lines." *The 2014 2nd International Conference on Systems and Informatics (ICSAI 2014)*, 2014, pp. 132–136. IEEE Xplore.

[3] Sun, Xiangwen, et al. "Harmonic Estimation Algorithm Based on ESPRIT and Linear Neural Network in Power System." *TELKOMNIKA (Telecommunication Computing Electronics and Control)*, vol. 14, no. 3A, Sept. 2016, pp. 47–55.

[4] Ortbandt, Claudius, et al. "Parameter Estimation in Electrical Power Systems Using Prony's Method." *Journal of Physics: Conference Series*, vol. 659, Nov. 2015, pp. 012–013.

[5] Khazaei, Javad, et al. "Distributed Prony Analysis for Real-World PMU Data." *Electric Power Systems Research*, vol. 133, Apr. 2016, pp. 113–120. Science Direct.

[6] Foyen, Sjur et al. "Prony's Method as a Tool for Power System Identification in Smart Grids." *2018 International Symposium on Power Electronics, Electrical Drives, Automation and Motion (SPEEDAM)*, 2018, pp. 562–569.

[7] Moravej, Z., et al. "Detection and Classification of Power Quality Disturbances Using Wavelet Transform and Support Vector Machines." *Electric Power Components and Systems*, vol. 38, no. 2, Dec. 2009, pp. 182–196. Taylor and Francis, NEJM.

[8] He, Haibo, et al. "Power Quality Disturbances Analysis Based on EDMRA Method." *International Journal of Electrical Power & Energy Systems*, vol. 31, no. 6, July 2009, pp. 258–268.

[9] Kordestani, Mojtaba, et al. "Harmonic Fault Diagnosis in Power Quality System Using Harmonic Wavelet." *IFAC-PapersOnLine*, vol. 50, no. 1, July 2017, pp. 13569–13574. ScienceDirect.

[10] Malla, Paul, et al. "Power System Fault Detection and Classification Using Wavelet Transform and Artificial Neural Networks." *Advances in Neural Networks – ISNN 2019*, edited by Huchuan Lu et al., Springer International Publishing, 2019, pp. 266–272. Springer Link.

[11] Jaya Bharata Reddy, Maddikara, et al. "A Multifunctional Real-Time Power Quality Monitoring System Using Stockwell Transform." *IET Science, Measurement Technology*, vol. 8, no. 4, July 2014, pp. 155–169. IEEE Xplore.

[12] Moravej, Z., et al. "New Combined S-Transform and Logistic Model Tree Technique for Recognition and Classification of Power Quality Disturbances." *Electric Power Components and Systems*, vol. 39, no. 1, Jan. 2011, pp. 80–98.

[13] Hasheminejad, S., et al. "Power Quality Disturbance Classification Using S-Transform and Hidden Markov Model." *Electric Power Components and Systems*, vol. 40, no. 10, July 2012, pp. 1160–1182.

[14] Biswal, Milan, and P. K. Dash. "Detection and Characterisation of Multiple Power Quality Disturbances with a Fast S-Transform and Decision Tree Based Classifier." *Digital Signal Processing*, vol. 23, no. 4, July 2013, pp. 1071–1083. ScienceDirect.

[15] Enshaee, Ali, and Parisa Enshaee. "A New S-Transform-Based Method for Identification of Power Quality Disturbances." *Arabian Journal for Science and Engineering*, vol. 43, no. 6, June 2018, pp. 2817–2832. Springer Link.

[16] Music, M., et al. "Integrated Power Quality Monitoring Systems in Smart Distribution Grids." *2012 IEEE International Energy Conference and Exhibition (ENERGYCON)*, 2012, pp. 501–506. IEEE Xplore.

[17] Chen, Cheng-I., et al. "Integrated Power-Quality Monitoring Mechanism for Microgrid." *IEEE Transactions on Smart Grid*, vol. 9, no. 6, Nov. 2018, pp. 6877–6885. IEEE Xplore.

[18] Bharath, B. S., and K. U. Vinayaka. "Investigation of Power Quality Disturbances in an Electric Arc Furnace." *2017 International Conference on Energy, Communication, Data Analytics and Soft Computing (ICECDS)*, 2017, pp. 2268–2273. IEEE Xplore.

[19] ABB DRIVES, Technical guide No. 6, Guide to harmonics with AC drives.

[20] EMTP 4.00.85, 2018®.

[21] MATLAB 2018a, The MathWorks, Natick, 2018.

Chapter 8

Microgrid control design with RES and electric vehicle integration

Shivam Jain

Indian Institute of Technology Roorkee, Roorkee, India

CONTENTS

8.1 INTRODUCTION

A microgrid is a small-scale energy distribution network, which enables the integration of a network of interconnected loads with a cluster of micro-sources within clearly defined electrical boundaries that can be operated in coordination and a controlled manner in both islanded and grid connected modes. It comprises of renewable sources of energy (e.g. solar, wind, fuel cell), energy storage devices (e.g. battery energy storage system, flywheel energy storage system) and non-renewable sources such as a diesel generator. In the grid connected mode of system operation, the frequency and voltage of the microgrid are dependent on the host grid; by contrast, in the islanded mode of operation, the microgrid operates independently and should be solely able to regulate voltage and frequency via the deployment of a secondary control mechanism. The power produced from the renewable sources of energy is highly dependent on the prevailing climatic conditions; therefore, it is uncontrollable and intermittent in nature. Further, the incorporation of renewable energy-based microsources as a replacement

DOI: 10.1201/9781003311195-8

of the conventional diesel generators leads to the reduction of inertia. In the islanded mode of system operation, fluctuations in frequency become particularly severe due to the small inertia and low time constants of micro-sources, which jeopardizes the stability and security of the system. The rapid frequency deviations in the islanded mode of microgrid operation leads to the need for a secondary control mechanism such that the frequency per-turbations are minimised and a reliable and good quality power is delivered at the consumer end.

The demand of electric power has skyrocketed due to various factors such as an increase in population, urbanisation and industrialisation. The increase in the demand for electric power cannot be satiated by the rapidly diminish-ing fossil fuel reserves, which are limited in nature and are a source of car-bon emissions, which lead to global warming and the greenhouse effect. Therefore, the usage of renewable sources of energy in the conventional power grid has become essential to provide clean, sustainable and environ-ment-friendly power to the consumers. However, the power generation from renewable sources of energy is variable and primarily dependent on the pre-vailing climate, which leads to a continuous imbalance between the genera-tion and demand of electric power. The stochastic nature of renewable sources of energy, massive reduction of inertia in the islanded mode of oper-ation, incessantly varying load demand, parametric uncertainty and the existence of nonlinearities such as governor dead-band and generation rate constraint causes fluctuations in frequency, which can even lead to power blackouts. Hence, the problem of load frequency control in a hybrid microgrid is one of the most challenging issues in the realm of power sys-tems and control engineering.

Hitherto, several techniques have been formulated in literature for the load frequency control of microgrids. The existing control techniques can be categorised into optimal control, nonlinear control, proportional-integral-derivative (PID)-based control, fractional order control, etc. The tuning of PID controller in [1] for a hybrid microgrid system is undertaken via the social spider optimization algorithm, wherein the integral time absolute error is considered as a performance index and various test scenarios are considered, such as load deviation, wind speed variations, etc. In [2], a non-linear sliding mode controller that also involves the incorporation of a hybrid version of wavelet mutation and sine cosine algorithm to ascertain the optimal values of controller coefficients is proposed for the load fre-quency control of a shipboard microgrid system. A fractional order cost function is employed with model predictive control for the load frequency control of an islanded microgrid in [3] and various simulation studies are conducted to demonstrate the efficacy over the conventional model predic-tive control algorithm. Further, the performance and robustness of an H_∞ controller and a iterative PID H_∞ controller designed via linear matrix inequalities is assessed in [4] for the problem of frequency regulation in a hybrid power system. It is observed that the latter controller exhibits robustness in the presence of parametric variations and also shows satisfactory disturbance rejection behaviour. The scaling factors of various PID-based classical fuzzy logic controllers are tuned via quasi-oppositional harmony search algorithm in [5] for the power control of a hybrid power system and the performance of the proposed controllers is investigated in the presence of wind perturbations and load demand. In [6], the tuning of a proportional integral controller is undertaken via the combination of an iterative particle swarm optimization algorithm and the concept of fuzzy logic for the regulation of frequency in the AC microgrid system. Several other techniques for the load frequency control of a hybrid microgrid system are elucidated in [7–11].

In this chapter, the mathematical modelling of a hybrid microgrid system is undertaken and various strategies for the load frequency control of microgrid are discussed. The components of a hybrid microgrid system are grouped into five different categories: controllable renewable sources (Biodiesel engine generator); non-controllable renewable energy sources (Solar, Wind); controllable non-renewable sources (Diesel generator); energy storage systems (Battery energy storage system, Ultracapacitor); and controllable loads (Plug-in hybrid electric vehicles). The working of these components is expounded and the corresponding linearized transfer function models are formulated. The block diagram of a hybrid microgrid system involving the aforementioned components is displayed. The integration of the microgrid (which is a building block of a smart grid) into the power system leads, however, to frequency deviations in a power system owing to the stochastic nature of renewable sources of energy, random fluctuations in load demand and parametric perturbations in the system. The deviations in frequency, which are more severe in the islanded mode of operation, necessitate the incorporation of a secondary control mechanism such that the power generation is balanced in consonance with the load demand and a continuous and superior quality power is delivered at the consumer end. Therefore, the load frequency control of microgrids is undertaken via the design of a secondary controller using linearized active disturbance rejection control (LADRC) technique.

The nonlinear active disturbance rejection control technique (ADRC) was proposed by Professor Han as an alternative to both classical and modern control techniques [12, 13]. It comprises of an extended state observer (ESO), which estimates not only the nominal states of the system, but also the combined effect of external disturbances and internal uncertainties via an additional state, which is commonly called the generalized disturbance or the total disturbance. Subsequently, the estimates of the states and the generalized disturbance are utilized in the state feedback-based control law, which is formulated such that the effect of generalized disturbance is attenuated in the overall controlled system. However, the nonlinear ADRC technique is complex and the number of tuning parameters in the nonlinear ADRC technique are large. To simplify the tuning procedure, the ESO and the estimated state feedback-based control law were linearized, leading to the development of the linearized active disturbance rejection control (LADRC) technique [14, 15]. The LADRC controller can be designed largely independent of the model-plant information and entails the tuning of only two parameters, namely observer bandwidth and controller bandwidth, which are directly linked to the performance of the system [16]. Several applications of ADRC techniques are given in [7, 17–21]. To demonstrate the load frequency control of a hybrid microgrid via linearized active disturbance rejection control technique, an elaborate case study and a comprehensive analysis is conducted. The sharing of power demand amongst the renewable sources of energy, energy storage systems and non-renewable sources in a hybrid microgrid, when a load perturbation occurs is shown via simulation results. The nature of load perturbations is considered as step variations to explain the load sharing mechanism, and random variations to mimic the practical load conditions prevailing in a particular area. An elaborate comparative analysis is undertaken, wherein the load frequency control of the hybrid microgrid system is shown in the absence and the presence of an LADRC controller. The simulation results establish the efficacy of the LADRC technique.

8.2 THE MATHEMATICAL MODELLING OF A MICROGRID

The constituents of a microgrid system can be categorised into renewable sources such as wind, biodiesel generator (BDeG), and solar, and non-renewable sources such as a diesel generator (DeG), energy storage systems such as ultracapacitor, and battery energy storage system and controllable loads like plug-in hybrid electric vehicles. In the given study, the BDeG and DeG are considered to be the controllable sources and the power generation from the controllable sources will be adjusted via a secondary control mechanism, so that the frequency fluctuations in the overall system converge to zero. The mathematical model of various microgrid components is given as follows (Figure 8.1).

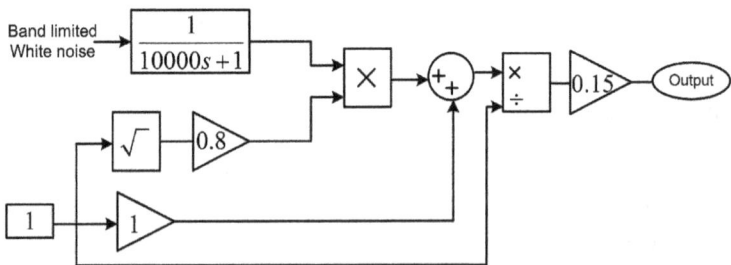

Figure 8.1 Solar farm.

8.2.1 Photovoltaic cell

A photovoltaic cell exploits the solar irradiation to create electron hole pairs in a *p-n* junction, leading to the generation of electric power. The solar energy is a renewable energy source and provides a sustainable alternative to the rapidly depleting non-renewable sources of energy. Moreover, the process of solar energy exploitation is noiseless and the photovoltaic cell has no moving parts. The transfer function of the photovoltaic cell can be expressed as [6]

$$G_{pv}(s) = \frac{\eta_{pv}}{1 + \gamma_{pv}s} \tag{8.1}$$

where η_{pv} represents the gain of te PV cell and γ_{pv} denotes the corresponding time constant. Figure 8.1 illustrates the model for the generation of randomly varying solar thermal power [22]. The random variations in solar thermal power are generated via the use of white noise block accompanied by a low pass filter [22]. The corresponding randomly variable solar power is shown in Figure 8.2.

8.2.2 Wind turbine generator

The wind turbine generator uses kinetic energy to turn the blades around a rotor, which further spins the prime mover of a generator for the production of electricity. The mechanical power produced via the wind turbine is a function of the wind velocity and power coefficient and is given as [23]

$$P_{wind} = \frac{1}{2} C_p v_w^3 \rho_a A_w \tag{8.2}$$

where, C_p represents the power coefficient, A_w denotes the area swept by wind turbine rotor, ρ_a signifies the density of air and v_w is the wind velocity. The formula for the computation of power coefficient is given as [24]

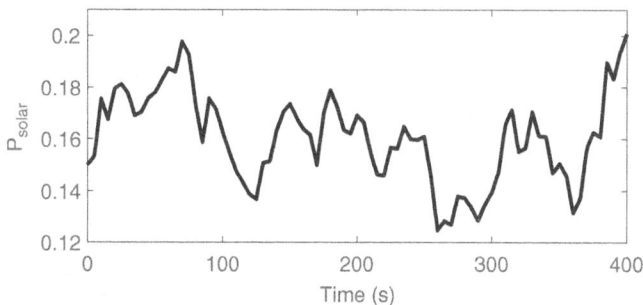

Figure 8.2 Random variation in solar power.

$$C_p = 0.0184\beta_b\left(-\frac{\omega_w R_r}{v_w} + 3\right) - (0.0167\beta_b - 0.44)\sin\left(\frac{\pi\left(-3 + \frac{\omega_w R_r}{v_w}\right)}{-0.3\beta_b + 15}\right) \quad (8.3)$$

Here, ω_w signifies angular velocity for the blades of turbine, R_r represents the wind turbine radius and β_b is the pitch angle of wind turbine blades. The dynamics of wind turbine generator are represented by the following transfer function

$$G_w(s) = \frac{\eta_w}{s\gamma_w + 1} \quad (8.4)$$

Figure 8.3 illustrates the model of a wind farm. The random variations in wind power can be generated via a wind farm by the use of white noise block accompanied with a low-pass filter [22, 25]. The corresponding randomly variable wind power is shown in Figure 8.4.

8.2.3 Biodiesel engine generator

The biodiesel fuel is a renewable alternative to the standard diesel fuel and is produced by transesterification, wherein vegetable oil and alcohol undergo chemical reaction in the presence of a catalyst. This leads to a reduction in harmful emissions of carbon monoxide, unburnt hydrocarbons and has improved lubrication properties. The linearized transfer function model of the Biodiesel engine generator (BDeG) without droop characteristics is given as [11]

$$G_{bd}(s) = \frac{K_v K_{bdeg}}{1 + (\gamma_v + \gamma_{bdeg})s + \gamma_v\gamma_{bdeg}s^2} \quad (8.5)$$

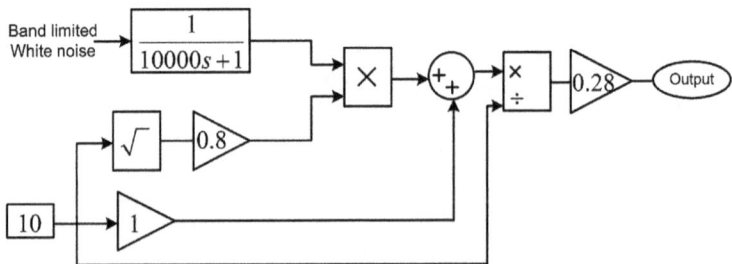

Figure 8.3 Wind farm.

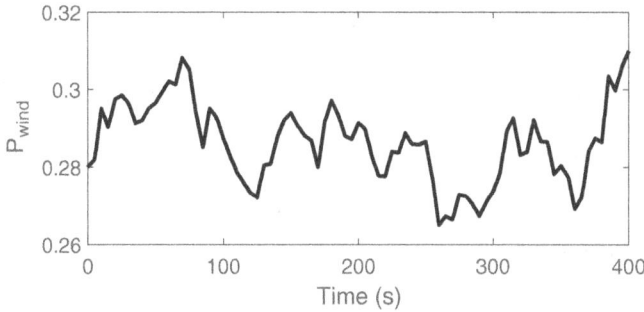

Figure 8.4 Randomly varying wind power.

where κ_{bdeg} and κ_v denote the gains of BDeG and valve, respectively. Further, γ_{bdeg} and γ_v represent the corresponding time constants.

8.2.4 Diesel generator

The diesel generator (DeG) is a non-renewable source for the production of electric power, which can be used as a standby when the power production from the renewables and other sources is insufficient to assuage the load demand. The mathematical model of DeG without droop characteristics is formulated as [7]

$$G_d(s) = \frac{1}{(\gamma_g s + 1)(\gamma_t s + 1)} \tag{8.6}$$

where γ_g and γ_t are the time constants of governor and turbine, respectively. Further, M is the inertia constant and D signifies the damping constant, respectively and R represents the droop characteristics.

8.2.5 Battery energy storage system

The energy storage systems help to smoothen the frequency fluctuations caused by the intermittent power generation from renewable sources of energy in order to provide reliable power to the consumers. The battery energy storage system (BESS) converts the chemical energy into electrical energy. During the period of high load demand, BESS supplies power to the grid, whereas when the production of power is greater than demand, it charges itself. The mathematical model of BESS is formulated as [22]

$$G_{be}(s) = \frac{\eta_b}{1 + s\gamma_b} \tag{8.7}$$

where η_b and γ_b signify the gain and time constants of BESS, respectively.

8.2.6 Plug-in electric vehicles

The plug-in hybrid electric vehicles (PhEV) act as a controllable load in a microgrid. During the peak demand of electric power, the PhEV acts in the vehicle to grid mode, wherein it supplies electric power to dampen the frequency fluctuations. On the other hand, during off-peak hours, the PhEV operates in the grid to vehicle mode and absorbs power to charge itself [26]. The mathematical model of the plug-in electric vehicles can be represented via the following transfer function [27]:

$$G_{\text{pev}}(s) = \frac{\eta_P}{1 + s\gamma_P} \tag{8.8}$$

where η_P and γ_P denote the corresponding gain and time constants, respectively.

8.2.7 Ultracapacitor

The ultracapacitor is an electric energy storage system which stores the energy in electric fields. The advantages of the ultracapacitor lie in its high power density and fast response times. It also helps in smoothing out the fluctuations in power output caused by the variation in environment-dependent renewable sources of energy. The mathematical model of the ultracapacitor is devised as [22, 28]

$$G_u(s) = \frac{\eta_{\text{uc}}}{1 + s\gamma_{\text{uc}}} \tag{8.9}$$

where η_{uc} represents the gain and γ_{uc} is the time constant of the ultracapacitor model.

The overall block diagram of the hybrid microgrid system comprising of all the aforementioned components is depicted in Figure 8.5. In the present work, the conventional diesel generator and biodiesel generator will be considered as the controllable components in the system. Further, P_l denotes the change in load demand and Δf represents the frequency deviations in the system. In what follows, a mathematical description of the linearized active disturbance rejection control technique will be given, which will be subsequently employed for the load frequency control of the hybrid microgrid system.

8.3 LINEARIZED ACTIVE DISTURBANCE REJECTION CONTROL

The linearized active disturbance rejection control technique can be employed for the design of a controller, when the mathematical model of the plant to be controlled is largely unknown to the control designer [14, 15].

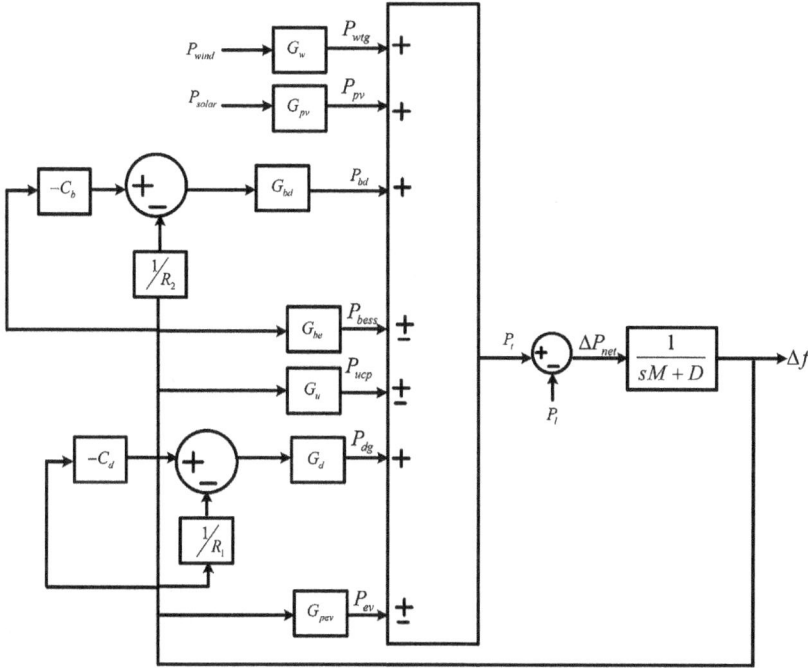

Figure 8.5 Block diagram of microgrid.

The block diagram of the LADRC technique is illustrated in Figure 8.6. The structure of LADRC can be categorised into the following parts:

- **Plant:** The plant denotes the system to be controlled via the application of the LADRC technique. The plant may be exposed to external disturbances in the environment and might also be affected by multifarious complicated uncertainties intrinsic to the plant such as nonlinearities, unmodelled dynamics, unknown terms, parametric uncertainty, etc.
- **Extended state observer:** The feedback control law requires complete information about the states of plant; however, in practice, all states may not be measurable. The task is then to formulate an observer that can estimate the unmeasured states as well as simultaneously obtain an estimation of the unknown disturbances affecting the plant. In the LADRC approach, it is the extended state observer (ESO) that is used to estimate the states of plant model as well as the generalized disturbance comprising of both internal uncertainties and external disturbance. By the careful tuning of gains in linear ESO, it is expected that the output of ESO can accurately recover the original states as well as the extended state of the system model.
- **Estimated state feedback control:** The last and a crucial part of an ADRC structure is the linear state feedback control law based on the

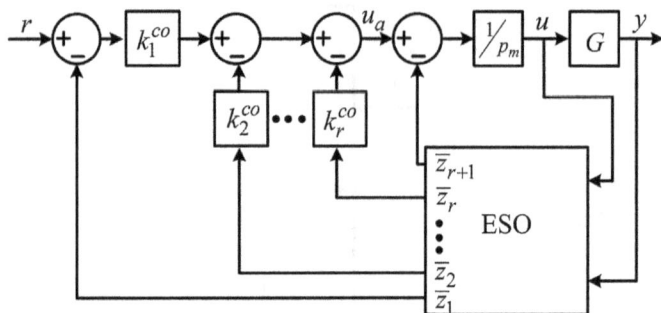

Figure 8.6 Structure of LADRC.

estimated states of plant. The purpose of linear state feedback control is to alleviate the effect of generalized disturbance by utilizing the estimate of extended state obtained via ESO for generation of an activating signal. The activation signal seeks to suppress and eliminate the adverse impact of generalized disturbance acting on plant. Finally, the output of plant $y(t)$ should also be able to efficiently track the reference signal by the application of ADRC approach.

In what follows, an elaborate description of the LADRC procedure is provided for a general nth-order linear system. Consider an nth-order nominal plant that can be represented by the following transfer function:

$$G(s) = \frac{Y(s)}{U(s)} = \frac{p_m s^m + \cdots + p_1 s + p_0}{s^n + q_{n-1} s^{n-1} + \cdots + q_1 s + q_0}; n \geq m \qquad (8.10)$$

However, the LADRC technique does not require complete information about the mathematical model of plant given in (8.10). Instead, the plant to be controlled can be designated by the following mathematical model:

$$y^{(r)}(t) = p_m u(t) + g(t) \qquad (8.11)$$

where r signifies the relative order of system model and is generally selected as $r = n - m$. Considering the system states as $\tilde{z}_1(t) = y(t)$, $\tilde{z}_2(t) = \dot{y}(t), \cdots$, $\tilde{z}_r(t) = y^{(r-1)}(t)$ and an additional virtual state variable as $\tilde{z}_{r+1}(t) = g(t)$, the extended state space representation of system is obtained as

$$\dot{\tilde{z}}(t) = A_{ob}\tilde{z}(t) + B_{ob}u(t) + E_{ob}h(t)$$
$$y(t) = C_{ob}\tilde{z}(t) \qquad (8.12)$$

where, $g(t)$ is assumed to be differentiable and $\dot{g}(t) = h(t)$. The state space matrices of the extended system are given as:

$$A_{ob} = \begin{bmatrix} 0 & 1 & 0 & \cdots & 0 & 0 \\ 0 & 0 & 1 & \cdots & 0 & 0 \\ \vdots & \vdots & \vdots & \ddots & \vdots & \vdots \\ 0 & 0 & 0 & \cdots & 0 & 1 \\ 0 & 0 & 0 & \cdots & 0 & 0 \end{bmatrix}_{(r+1) \times (r+1)}, B_{ob} = \begin{bmatrix} 0 \\ 0 \\ \vdots \\ p_m \\ 0 \end{bmatrix}_{(r+1) \times 1}$$

$$C_{ob} = \begin{bmatrix} 1 & 0 & \cdots & 0 & 0 \end{bmatrix}_{1 \times (r+1)}, E_{ob} = \begin{bmatrix} 0 & 0 & \cdots & 0 & 1 \end{bmatrix}^T_{(r+1) \times 1}$$

(8.13)

where the dimensionality of the matrices A_{ob}, B_{ob}, C_{ob} and E_{ob} is $(r + 1) \times (r + 1)$, $(r + 1) \times 1$, $1 \times (r + 1)$ and $(r + 1) \times 1$, respectively. Next, a linear extended state observer is constructed from the process output and input data for undertaking the estimation of plant output, its derivatives and the generalized disturbance. The mathematical model of the ESO is formulated as

$$\dot{\bar{z}}(t) = A_{ob}\bar{z}(t) + B_{ob}u(t) + L_{ob}\left(y(t) - \bar{y}(t)\right)$$
$$\bar{y}(t) = C_{ob}\bar{z}(t)$$

(8.14)

where, $\bar{z}(t) = \begin{bmatrix} \bar{z}_1(t) & \bar{z}_2(t) & \cdots & \bar{z}_r(t) & \bar{z}_{r+1}(t) \end{bmatrix}^T$ are the observer states, $\bar{y}(t)$ represents the output of ESO and $L_{ob} = \begin{bmatrix} \beta_1^{ob} & \beta_2^{ob} & \cdots & \beta_{r+1}^{ob} \end{bmatrix}^T$ signifies the observer gain matrix. Most significantly, the state $\bar{z}_{r+1}(t)$ estimates the generalized disturbance $g(t)$. The observer states $\bar{z}_1(t), \bar{z}_2(t), \cdots, \bar{z}_r(t)$ will approximate the system states and $\bar{z}_{r+1}(t)$ tracks $g(t)$, when $(A_{ob} - L_{ob}C_{ob})$ is asymptotically stable. The estimate of states as well as generalized disturbance will be subsequently employed in the control law to attenuate the disturbance and ensure an effective control of the system.

The control rule is chosen as

$$u(t) = \frac{u_a(t) - \bar{z}_{r+1}(t)}{p_m}$$

(8.15)

Substituting the expression of control rule from (8.15) into (8.11), the resulting closed loop plant is obtained as

$$y^{(r)}(t) = p_m\left(\frac{u_a(t) - \bar{z}_{r+1}(t)}{p_m}\right) + g(t)$$
$$= u_a(t) - \bar{z}_{r+1}(t) + g(t)$$

(8.16)

Assuming that the observer is designed in such a way that the generalized disturbance is timely estimated in an accurate manner, (i.e., $\bar{z}_{r+1}(t) \approx g(t)$), Equation (8.16) reduces to

$$y^{(r)}(t) = u_a(t) \tag{8.17}$$

It can be discerned that the estimated disturbance signal has been employed in control law for the generation of an activation signal that proceeds to cancel the effect of generalized disturbance itself. Moreover, the original plant in (8.10) has been reduced to a rth-order integral plant in (8.17). The signal $u_a(t)$ is selected based on the estimated state feedback mechanism as follows:

$$u_a(t) = k_1^{co}\left(r(t) - \bar{z}_1(t)\right) - k_2^{co}\bar{z}_2(t) - \cdots - k_r^{co}\bar{z}_r(t) \tag{8.18}$$

Using (8.18), Equation (8.15) can be re-written as

$$u = \frac{k_1^{co}\left(r(t) - \bar{z}_1(t)\right) - k_2^{co}\bar{z}_2(t) - \cdots - k_r^{co}\bar{z}_r(t) - \bar{z}_{r+1}(t)}{p_m} \tag{8.19}$$

$$= K_{co}\left(\tilde{r}(t) - \bar{z}(t)\right)$$

where $\tilde{r}(t) = \begin{bmatrix} r(t) & 0 & \cdots & 0 \end{bmatrix}^T$ denotes the reference signal matrix of system and $K_{co} = \dfrac{1}{p_m}\begin{bmatrix} k_1^{co} & k_2^{co} & \cdots & k_r^{co} & 1 \end{bmatrix}$ represents the controller gain matrix. Therefore, LADRC controller design entails tuning of the controller gain matrix K_{co} and the observer gain matrix L_{ob}. Using the bandwidth parameterization approach, the procedure for evaluation of the two gains, i.e., K_{co} and L_{ob}, can be reduced to the computation of two bandwidths of the system, i.e., controller bandwidth (ω_{co}) and observer bandwidth (ω_{ob}) [16]. The choice of bandwidth also reflects multiple trade-offs in control system design. A higher value of observer bandwidth leads to a better state estimation ability, and a higher value of controller bandwidth gives rise to a quicker speed of response. However, faster system response also brings about higher overshoot and reduction of system stability. Meanwhile, higher observer bandwidth also makes the system prone to high-frequency noise. Therefore, selection of the bandwidths is intricately linked to the overall system performance. For the placement of all observer poles at $-\omega_{ob}$, Ackermann's formula can be used for the computation of observer gain matrix. Therefore, the observer gain matrix is obtained as

$$L_{ob} = \psi(A_{ob}) \begin{bmatrix} C_{ob} \\ C_{ob}A_{ob} \\ \vdots \\ C_{ob}A_{ob}^r \end{bmatrix}^{-1} \begin{bmatrix} 0 \\ 0 \\ \vdots \\ 1 \end{bmatrix} \tag{8.20}$$

where $\psi(A_{ob})$ is obtained by evaluating $\psi(s) = (s + \omega_{ob})^{r+1}$ at $s = A_{ob}$. Further simplification of (8.20) yields the observer gains as

$$\beta_j^{ob} = \binom{r+1}{j}\omega_{ob}^j, \quad j = \{1,2,\cdots,r+1\} \tag{8.21}$$

In a similar manner, the characteristic polynomial of estimated state feedback controller can be given as

$$\phi_k(s) = |sI - A_{ob} + B_{ob}K_{co}$$
$$= s\left(s^r + k_r^{co}s^{r-1} + k_{r-1}^{co}s^{r-2} + \cdots + k_2^{co}s + k_1^{co}\right) \tag{8.22}$$

Placement of r controller poles at $-\omega_{co}$ yields the desired characteristic polynomial as

$$\phi_d(s) = s(s + \omega_{co})^r \tag{8.23}$$

Comparing (8.22) and (8.23), the controller gains are obtained as

$$k_j^{co} = \binom{r}{j-1}\omega_{co}^{r-(j-1)}, \quad j = \{1,2,\cdots,r\} \tag{8.24}$$

Hence, observer bandwidth and controller bandwidth are the two tuning parameters for the design of the LADRC controller for linear systems. The LADRC controller design does not depend on the model of plant, except for the relative order and the high-frequency gain of the plant. Furthermore, the LADRC controller has a fixed structure and involves the tuning of only two parameters, making it easily envisaged by the practitioners of control theory.

8.4 SIMULATION RESULTS

In this section, the linearized active disturbance rejection control technique will be employed for the controller design of the controllable elements in the hybrid microgrid system. The parameters of the hybrid microgrid system are considered as [7, 11, 27]

$$\eta_w = 1, \gamma_w = 1.5, \eta_{pv} = 1, \gamma_{pv} = 1.8, \kappa_\nu = 1, \kappa_{bdeg} = 1, \gamma_v = 0.05, \gamma_{bdeg} = 0.5,$$
$$\eta_b = 0.0033, \gamma_b = 0.1, \eta_P = 1, \gamma_P = 0.2, D = 0.015, M = 0.1667, \tag{8.25}$$
$$\gamma_g = 0.08, \gamma_t = 0.4, \eta_{uc} = 0.7, \gamma_{uc} = 0.9, R_1 = 3$$

Using the parameters from (8.25), the overall transfer function of the controllable elements, i.e., biodiesel engine generator (BDeG) and diesel generator (DeG), are computed as

$$G_1(s) = \frac{1}{0.004168s^3 + 0.09206s^2 + 0.1749s + 0.4317}$$

$$G_2(s) = \frac{1}{0.005333s^3 + 0.08048s^2 + 0.1739s + 0.3483} \quad (8.26)$$

The state space matrices of the extended state observer for BDeG and DeG are obtained as

$$A_{ob1} = \begin{bmatrix} 0 & 1 & 0 & 0 \\ 0 & 0 & 1 & 0 \\ 0 & 0 & 0 & 1 \\ 0 & 0 & 0 & 0 \end{bmatrix}, \quad B_{ob1} = \begin{bmatrix} 0 \\ 0 \\ 239.9 \\ 0 \end{bmatrix}, \quad C_{ob1} = \begin{bmatrix} 1 & 0 & 0 & 0 \end{bmatrix} \quad (8.27)$$

$$A_{ob2} = \begin{bmatrix} 0 & 1 & 0 & 0 \\ 0 & 0 & 1 & 0 \\ 0 & 0 & 0 & 1 \\ 0 & 0 & 0 & 0 \end{bmatrix}, \quad B_{ob2} = \begin{bmatrix} 0 \\ 0 \\ 187.5 \\ 0 \end{bmatrix}, \quad C_{ob2} = \begin{bmatrix} 1 & 0 & 0 & 0 \end{bmatrix} \quad (8.28)$$

Using the formulae in (8.21) and (8.24), the controller (ω_{co} = 40) and observer gain matrices (ω_{ob} = 80) for BdeG and DeG are obtained as

$$K_{co2} = \begin{bmatrix} 341.3333 & 25.6000 & 0.6400 & 0.0053 \end{bmatrix}$$

$$L_{ob2} = \begin{bmatrix} 320 & 38400 & 2048000 & 40960000 \end{bmatrix}^T \quad (8.29)$$

An extensive case study is considered to study the effectiveness of the LADRC control technique to dampen the frequency deviations caused due to randomly varying load demand and due to the introduction of low-inertia renewable sources of energy. The scenarios are considered as given below.

8.4.1 Scenario I

In Scenario 1, the components of the microgrid to be considered are the photovoltaic cell, the wind turbine generator, the battery energy storage system, the ultracapacitor and the diesel generator. Using the formulae in (8.21) and (8.24), the controller (ω_{co} = 40) and observer gain matrices (ω_{ob} = 80) for DeG are obtained as

$$K_{co2} = \begin{bmatrix} 341.3333 & 25.6000 & 0.6400 & 0.0053 \end{bmatrix}$$

$$L_{ob2} = \begin{bmatrix} 320 & 38400 & 2048000 & 40960000 \end{bmatrix}^T \quad (8.30)$$

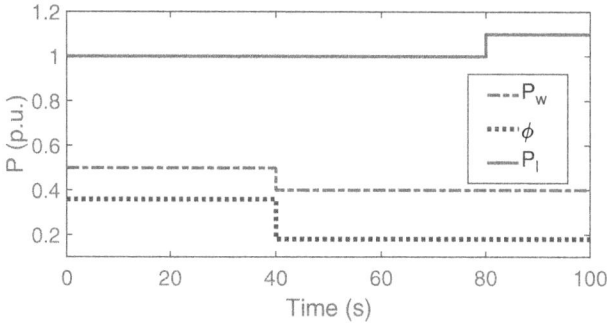

Figure 8.7 Input renewable and load power fluctuations for scenario 1.

The step perturbations are considered in the renewable sources of energy and the load. The input wind power is considered to be 0.5 p.u. for $t < 40$ s and 0.4 p.u. for $t \geq 40$ s. The solar input is taken as 0.36 p.u. for $t < 40$ s and 0.18 p.u. for $t \geq 40$ s. The load demand is considered to be 1 p.u for $t < 40$ s and 1.1 p.u for $t \geq 40$ s. The corresponding plot is shown in Figure 8.7.

To demonstrate the need of the controller for countering the deviations in frequency, an analysis is conducted, wherein the frequency deviations in the presence of renewable sources of energy are shown in the absence of the LADRC controller as well as in the presence of the LADRC controller.

The corresponding responses are shown in Figure 8.8. It can be observed here that in the absence of the controller, the hybrid microgrid system exhibits large overshoots and undershoots, and steady state frequency deviations at the onset of perturbations in wind and solar at $t = 40$ s and load at $t = 80$ s, respectively. On the other hand, in the presence of the LADRC controller, the fluctuations in frequency are quickly attenuated accompanied by minimal overshoot and undershoot as compared to the case when controller design is not considered. Therefore, the design of a suitable secondary controller is vital in ensuring effective disturbance rejection.

The power sharing among different components of the microgrid is depicted in Figure 8.9. At time $t > 40$ s, the power output from photovoltaic cell and wind turbine generator reduces, however, the load power demand remains does not change as compared to the value when $t < 40$ s. Since the power generation from renewable sources has reduced, therefore, the controllable power source, i.e. the diesel generator, should increase generation, since the load demand is unchanged. It can be seen from Figure 8.9f that at $t = 40$ s, the power output from DeG increases in order to meet the required load demand. Further, at $t = 80$ s, the load demand increases; however, the power generation from renewable sources of energy has remained constant. To meet the increased load demand, the power generation from DeG increases further. Hence, it can be ascertained that the controllable sources

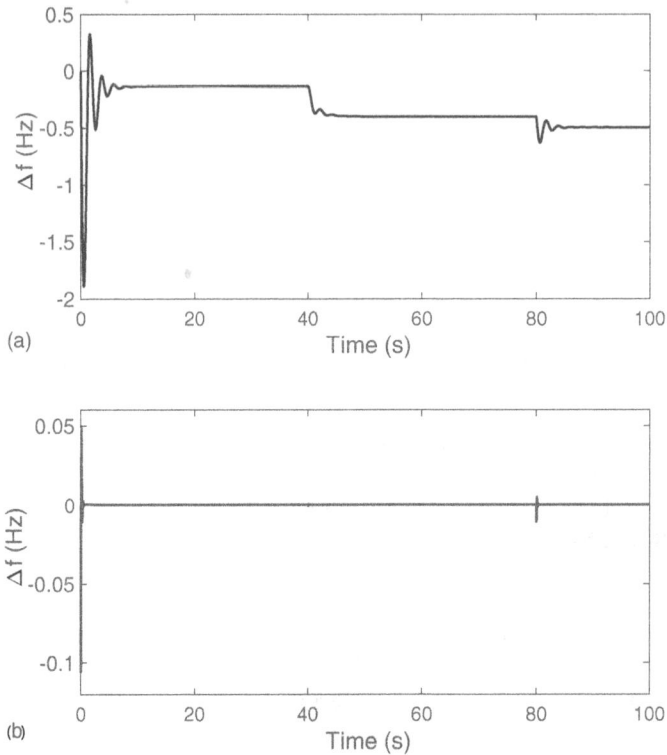

(a)

(b)

Figure 8.8 Frequency fluctuations comparison in scenario I.

of energy adjust their power generation in accordance with the changes in renewable power generation and the variation in load demand.

8.4.2 Scenario 2

In Scenario 2, the components of the microgrid to be considered are photovoltaic cell, wind turbine generator, battery energy storage system, electric vehicle, conventional DeG and biodiesel generator. Using the formulae in (8.21) and (8.24), the controller (ω_{co} = 5) and observer gain matrices (ω_{ob} = 12) for BDeG are obtained as

$$
\begin{aligned}
K_{co1} &= \begin{bmatrix} 0.5211 & 0.3126 & 0.0625 & 0.0042 \end{bmatrix} \\
L_{ob1} &= \begin{bmatrix} 48 & 864 & 6912 & 20736 \end{bmatrix}^{T}
\end{aligned}
\tag{8.31}
$$

In this scenario, the random perturbations are considered in the renewable sources of energy and load to mimic the practical scenario, wherein the perturbations are random in nature. The random fluctuations in solar and

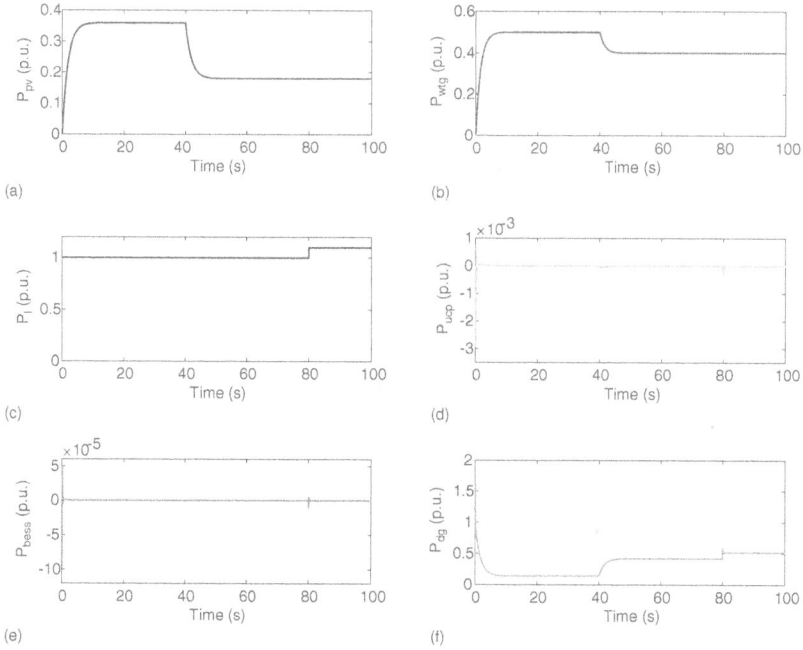

(a)
(b)
(c)
(d)
(e)
(f)

Figure 8.9 Power output of various components of the microgrid in scenario 1.

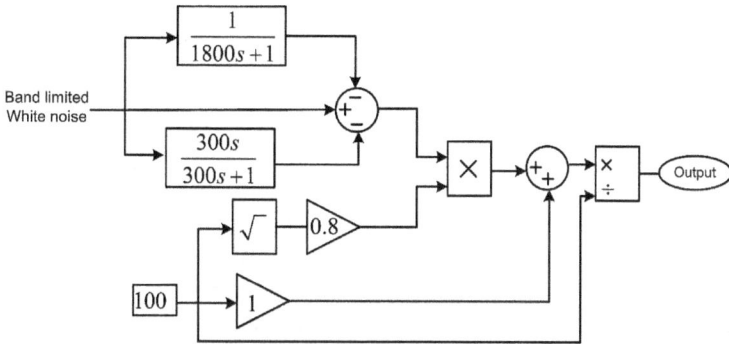

Figure 8.10 Model for variable load power.

wind to be considered in this scenario are depicted in Figures 8.2 and Figure 8.4, respectively. The model for generation of random variations in load is shown in Figure 8.10 and the corresponding randomly varying load demand is illustrated in Figure 8.11.

In the presence of randomly varying solar, wind and load demand, the frequency fluctuation response is ascertained and the corresponding response is shown in Figure 8.12. It can be observed that in the presence of the

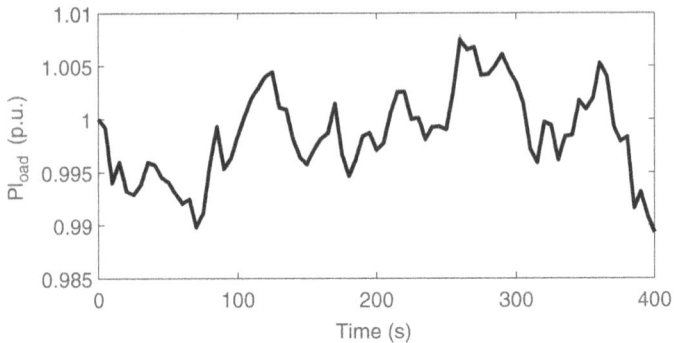

Figure 8.11 Load power deviation in scenario 2.

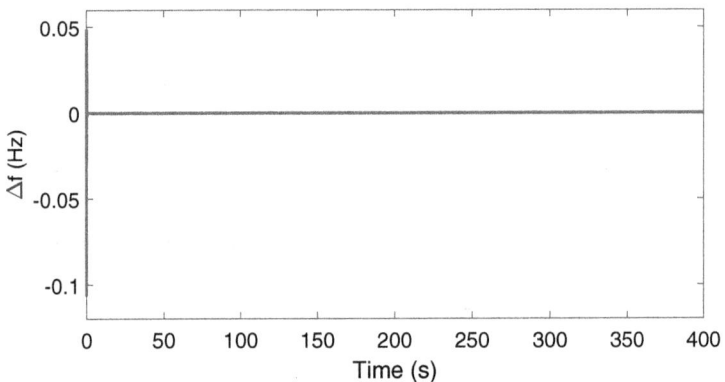

Figure 8.12 Frequency deviation in scenario 2.

LADRC controller, the fluctuations in frequency are quickly attenuated, accompanied by minimal overshoot and undershoot.

Hence, the aforementioned study validates the efficacy of the LADRC technique in eliminating the frequency fluctuations caused by the presence of renewable sources of energy and the randomly varying load demand.

8.5 CONCLUSION

In this chapter, the linearized active disturbance rejection control technique, which is largely model-independent in nature, is applied for the load frequency control of a hybrid microgrid system in the presence of step perturbations as well as random variations in solar, wind and load demand. The LADRC technique has only two tuning parameters and can be used to design a controller, even when the model plant information is largely unknown to the control designer. It is observed that the incorporation of the

LADRC controller enables the elimination of the frequency fluctuations in a hybrid microgrid system. In future, the effect of communication delay can be considered while applying the LADRC technique for the load frequency control of a hybrid microgrid and other complex power systems.

REFERENCES

1. A. A. El-Fergany and M. A. El-Hameed, "Efficient frequency controllers for autonomous two-area hybrid microgrid system using social-spider optimiser," *IET Generation, Transmission & Distribution*, vol. 11, no. 3, pp. 637–648, 2017. [Online]. Available: https://ietresearch.onlinelibrary.wiley.com/doi/abs/10.1049/iet-gtd.2016.0455

2. M.-H. Khooban, M. Gheisarnejad, N. Vafamand, M. Jafari, S. Mobayen, T. Dragicevic, and J. Boudjadar, "Robust frequency regulation in mobile microgrids: Hil implementation," *IEEE Systems Journal*, vol. 13, no. 4, pp. 4281–4291, 2019.

3. M.-R. Chen, G.-Q. Zeng, Y.-X. Dai, K.-D. Lu, and D.-Q. Bi, "Fractional-order model predictive frequency control of an islanded microgrid," *Energies*, vol. 12, no. 1, 2019. [Online]. Available: https://www.mdpi.com/1996-1073/12/1/84

4. S. K. Pandey, N. Kishor, and S. R. Mohanty, "Frequency regulation in hybrid power system using iterative proportional-integral-derivative h_∞ controller," *Electric Power Components and Systems*, vol. 42, no. 2, pp. 132–148, 2014. [Online]. Available: https://doi.org/10.1080/15325008.2013.846438

5. S. Ganguly, T. Mahto, and V. Mukherjee, "Integrated frequency and power control of an isolated hybrid power system considering scaling factor based fuzzy classical controller," *Swarm and Evolutionary Computation*, vol. 32, pp. 184–201, 2017. [Online]. Available: https://www.sciencedirect.com/science/article/pii/S221065021630181X

6. H. Bevrani, F. Habibi, P. Babahajyani, M. Watanabe, and Y. Mitani, "Intelligent frequency control in an ac microgrid: Online pso-based fuzzy tuning approach," *IEEE Transactions on Smart Grid*, vol. 3, no. 4, pp. 1935–1944, 2012.

7. S. Jain and Y. V. Hote, "Generalized active disturbance rejection controller design: Application to hybrid microgrid with communication delay," *International Journal of Electrical Power & Energy Systems*, vol. 132, p. 107166, 2021. [Online]. Available: https://www.sciencedirect.com/science/article/pii/S0142061521004051

8. D. Oliveira, A. Zambroni de Souza, M. Santos, A. Almeida, B. Lopes, and O. Saavedra, "A fuzzy-based approach for microgrids islanded operation," *Electric Power Systems Research*, vol. 149, pp. 178–189, 2017. [Online]. Available: http://www.sciencedirect.com/science/article/pii/S0378779617301669

9. L. Yin, T. Yu, B. Yang, and X. Zhang, "Adaptive deep dynamic programming for integrated frequency control of multi-area multi-microgrid systems," *Neurocomputing*, vol. 344, pp. 49–60, 2019, neural learning in life system and energy system. [Online]. Available: http://www.sciencedirect.com/science/article/pii/S0925231219301663

10. I. Pan and S. Das, "Kriging based surrogate modeling for fractional order control of microgrids," *IEEE Transactions on Smart Grid*, vol. 6, no. 1, pp. 36–44, 2015.

11. A. K. Barik and D. C. Das, "Expeditious frequency control of solar photovoltaic/biogas/biodiesel generator based isolated renewable microgrid using grasshopper optimisation algorithm," *IET Renewable Power Generation*, vol. 12, no. 14, pp. 1659–1667, 2018, [Online]. Available: https://ietresearch.onlinelibrary.wiley.com/doi/abs/10.1049/iet-rpg.2018.5196

12. J. Han, "From pid to active disturbance rejection control," *IEEE Transactions on Industrial Electronics*, vol. 56, no. 3, pp. 900–906, 2009.

13. B. Guo and Z. Zhao, *Active Disturbance Rejection Control for Nonlinear Systems*. Singapore: John Wiley & Sons, Ltd, 2016. https://onlinelibrary.wiley.com/doi/book/10.1002/9781119239932

14. Z. Gao, "Active disturbance rejection control: a paradigm shift in feedback control system design," in *2006 American Control Conference*, June 2006.

15. W. Tan and C. Fu, "Linear active disturbance-rejection control: Analysis and tuning via IMC," *IEEE Transactions on Industrial Electronics*, vol. 63, no. 4, pp. 2350–2359, April 2016.

16. Zhiqiang Gao, "Scaling and bandwidth-parameterization based controller tuning," in *Proceedings of the 2003 American Control Conference, 2003.*, vol. 6, 2003, pp. 4989–4996.

17. S. Jain and Y. V. Hote, "Generalized active disturbance rejection controller for load frequency control in power systems," *IEEE Control Systems Letters*, vol. 4, no. 1, pp. 73–78, 2019.

18. S. Jain and Y. V. Hote, "Design of generalised active disturbance rejection control for delayed systems: an application to load frequency control," *International Journal of Control*, vol. 0, pp. 1–15, 2020. [Online]. Available: https://doi.org/10.1080/00207179.2020.1752940

19. B. Ahi and A. Nobakhti, "Hardware implementation of an ADRC controller on a gimbal mechanism," *IEEE Transactions on Control Systems Technology*, vol. 26, no. 6, pp. 2268–2275, 2018.

20. Q. Zheng, L. Dong, D. H. Lee, and Z. Gao, "Active disturbance rejection control for MEMS gyroscopes," *IEEE Transactions on Control Systems Technology*, vol. 17, no. 6, pp. 1432–1438, 2009.

21. X. Chang, Y. Li, W. Zhang, N. Wang, and W. Xue, "Active disturbance rejection control for a flywheel energy storage system," *IEEE Transactions on Industrial Electronics*, vol. 62, no. 2, pp. 991–1001, 2015.

22. D. C. Das, A. Roy, and N. Sinha, "Ga based frequency controller for solar thermal–diesel–wind hybrid energy generation/energy storage system," *International Journal of Electrical Power & Energy Systems*, vol. 43, no. 1, pp. 262–279, 2012. [Online]. Available: http://www.sciencedirect.com/science/article/pii/S0142061512002153

23. G. Shankar and V. Mukherjee, "Load frequency control of an autonomous hybrid power system by quasi-oppositional harmony search algorithm," *International Journal of Electrical Power & Energy Systems*, vol. 78, pp. 715–734, 2016. [Online]. Available: https://www.sciencedirect.com/science/article/pii/S0142061515005220

24. D. Lee and L. Wang, "Small-signal stability analysis of an autonomous hybrid renewable energy power generation/energy storage system part i: Time-domain simulations," *IEEE Transactions on Energy Conversion*, vol. 23, no. 1, pp. 311–320, 2008.

25. M. Matsubara, G. Fujita, T. Shinji, T. Sekine, A. Akisawa, T. Kashiwagi, and R. Yokoyama, "Supply and demand control of dispersed type power sources in micro grid," in *Proceedings of the 13th International Conference on, Intelligent Systems Application to Power Systems*, 2005, pp. 67–72.
26. A. Latif, A. Pramanik, D. C. Das, I. Hussain, and S. Ranjan, "Plug in hybrid vehicle-wind-diesel autonomous hybrid power system: frequency control using fa and csa optimized controller," *International Journal of System Assurance Engineering and Management*, vol. 9, no. 5, pp. 1147–1158, 2018.
27. A. Latif, D. C. Das, S. Ranjan, and A. K. Barik, "Comparative performance evaluation of wca-optimised non-integer controller employed with wpg–dspg–phev based isolated two-area interconnected microgrid system," *IET Renewable Power Generation*, vol. 13, no. 5, pp. 725–736, 2019. [Online]. Available: https://ietresearch.onlinelibrary.wiley.com/doi/abs/10.1049/iet-rpg.2018.5419
28. M. Faisal, M. A. Hannan, P. J. Ker, A. Hussain, M. B. Mansor, and F. Blaabjerg, "Review of energy storage system technologies in microgrid applications: Issues and challenges," *IEEE Access*, vol. 6, pp. 35, 143–135, 164, 2018.

Chapter 9

A smart grid with renewable energy sources, e-vehicles, and storage systems

Operational and economic aspects

Felipe Sabadini

RWTH Aachen University, Aachen, Germany

Reinhard Madlener

RWTH Aachen University, Aachen, Germany

Norwegian University of Science and Technology (NTNU), Trondheim, Norway

CONTENTS

DOI: 10.1201/9781003311195-9

9.1 INTRODUCTION

In a scenario to mitigate greenhouse gas (GHG) emissions, renewable energies are becoming increasingly important with every year that passes, and variable renewable energy (VRE), such as solar photovoltaics (PV) and wind, will play a vital role in a sustainable future. However, such intermittent power generation, originating mainly from solar PV and wind, is often unable to match demand and can thus cause a problem with balancing. One often-used remedy for this challenge is to use battery energy storage systems (BESS), as these can store energy and discharge it at short notice when demand occurs. With many different variables, such as intermittent energy-influencing factors, storage units, the communication between several energy technologies and the control of a complex operation depends on the development of the so-called "smart grid". A smart grid is a modernized electrical grid that uses analog or digital information and communications technology. It incorporates various technologies – such as the internet of things, power control strategies, and end-user applications [1].

Traditionally, the supply side of conventional power systems offered flexibility by unilaterally adjusting generation to meet demand. Since the demand side was essentially unresponsive, it provided very little versatility. Currently, the flexibility comes from sources that can ramp up or down rapidly, and have a relatively low operating cost level, and quick start-up and shutdown times. In the present day, storage hydro generators and open-cycle gas turbines are considered to be among the most flexible conventional power generation types.

However, the commercial availability and mass market diffusion of technologies, such as BESS and demand-side flexibility, are able not only to expand flexibility options, but can also help to improve system reliability. We expect the system to change towards a bilateral flow which increasingly uses BESS connected to a VER plant and storing its excess renewable electricity. The demand side can also provide significant flexibility through either the direct or indirect electrification of end-use sectors. If the load is well managed, BESS can also help to increase grid flexibility and reduce network congestion. Figure 9.1 shows these changes in the direction of a flexible framework with the participation of VRE.

Furthermore, when we consider that governments will increase their share of renewables in the future, the integration of VRE poses specific challenges as the share of (intermittent) power generation rises – in essence, maintaining the balance of supply and demand becomes an increasing challenge [2]. As the share of renewables in the power mix increases, more flexible possibilities are needed to maximize the value of renewable energy, especially solar and wind. Enhancing the absorptive capacity of the system with respect to VRE will enable the provision of cost-effective renewable energy while improving power quality and reliability. In addition, it helps to reduce (shave) load peaks whilst simultaneously cutting greenhouse gas emissions [2].

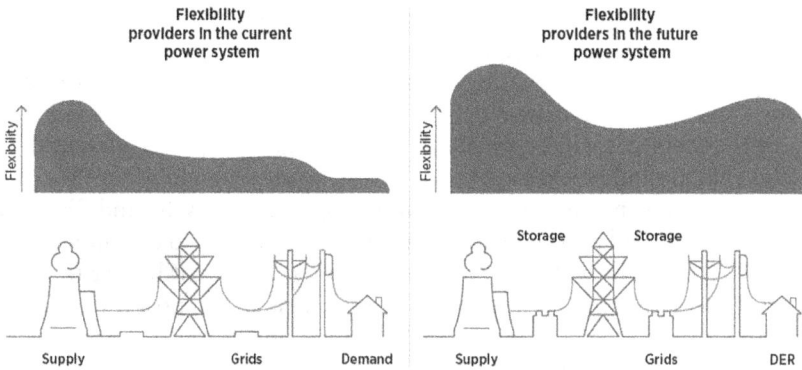

Figure 9.1 Transformation of the power systems framework to a more flexible and decentralized system. (Source: IRENA [2]).

Next, we discuss in more detail the legislation, policies and market opportunities concerning the incorporation of vehicle-to-grid (V2G) and stationary battery storage capacity as an integral part of smart electricity grid systems and their challenges.

9.2 OPERATIONAL PARAMETERS: GUIDELINES AND STANDARDIZATION

9.2.1 Governance and legislation

The unclear role of BESS/V2G in existing legal and regulatory structures is a vital issue. Although the rule is that power markets are usually designed for power generation units, current and future regulatory frameworks regarding energy storage ought to be adapted to the specific needs of different regions and energy markets, respectively, while also being adapted for storage units.

Regulation agencies in many countries across the world are under pressure to maintain an optimal level of service at an affordable price for their taxpayers. Changing existing policies and regulations may cause disturbance to a situation of "business as usual"; however, the introduction of V2G and BESS can help to enhance several societal benefits, such as grid efficiency, revenue to car owners from their participation in the electricity generation market, new business models for companies, and the increased integration of renewable energy into the power system [3, 4].

In addition to the lack of classification and recognition of these technologies in most of the world's energy markets, a lack of a regulatory definition for aggregators, immature market structures, and the grid itself not being smart enough are among the other issues discussed in the literature.

According to Noel et al. [5], V2G, for instance, has to date only been the focus of research and demonstration programmes, and the technology has yet to be fully commercialized. As a consequence, no law explicitly regulating V2G. exists in the majority of countries. However, to face emerging problems with more fluctuating clean energy in the electricity sector, electricity supply regulations will need to change [6]. As a result, it is expected that V2G operators will be forced to adhere to general power market regulations that are unfavourable to this specific business model [7].

Even though some markets allow for the operation of V2G and BESS, the authors have found that there are no incentives available to encourage commercial aggregators to participate in the market, making it harder to establish V2G without any financial incentive [4]. One of the possible solutions that benefits the economic viability of batteries operators, in general, might be asymmetrical bids with preferable market products aiming at commercial aggregators. To implement this idea, the V2G/BESS aggregators can have separate auctions that require asymmetric reserves that would benefit the standard nature of battery systems (e.g. charging/discharging), such as up- and down-regulation reserves. Another specific barrier to V2G operations is the number of electric vehicles (EVs) necessary to participate in the market, which is normally the necessary amount of cars to get to achieve a 1MW capacity. The legislation can have lower requirements for these actors; such a move would not only benefit V2G projects, but all different types of BESS, since the initial investment in storage systems is a significant barrier to this operation.

Although we use BESS as a general term, it is important to highlight that V2G and stationary batteries have many differences in the way they operate, thereby also facing specific barriers. One of the many examples we can use is that of different, operation-dependent battery degradation costs, since the EV battery must bear both normal driving degradation and V2G operation; V2G also faces additional operational constraints, such as driving patterns and keeping the battery charged at a minimum level so that a driver can still drive the EV at any time if needed; in addition, the transaction costs are significantly larger due to searching for, and contracting with, the owners of a larger car fleet.

Even the current design of EVs is currently unsuited for the proposed V2G infrastructure, and technical changes are expected from car manufacturers. In this scenario, V2G implementation will complicate the EV technology still further, resulting in increased vehicle and infrastructure costs; and the price is already a barrier to EVs, being significantly more expensive than for conventional cars, e.g. IRENA [2].

When focusing on the charging barriers, we find significant problems; V2G is hampered by the lack of charging stations and the fact that only small numbers of EV models support bidirectional charging, which is necessary for V2G to operate. Furthermore, bidirectional chargers must be included in the charging infrastructure, especially for "standard" charging in parking areas where EVs are parked the majority of the time [7].

9.2.2 Market opportunities

Many different opportunities arise when discussing the exploration of BESS. At the most simple level of analysis, an EV can help a household to increase the amount of electricity consumed that is produced from the rooftop solar PV system while not being so prohibitive in terms of driving freedom. Different studies show that the use of a vehicle-to-home system (V2H) can reduce the levelized cost of electricity (LCOE) of PV-battery domestic systems. Angenendt et al. [8], in addition, finds that V2H also reduces the charging costs of the vehicle, saving up to 11% [9]. However, given the personal choices of each household and how they deeply influence the final economic result, in this section we consider strategies to provide ancillary services to the grid. This happens because the results can be generalized and because these services are safer in terms of battery degradation, as frequency control allows for shallow charge/discharge cycles rather than deep depth of discharge (DoD).

Before exploring different market opportunities for both V2G and BESS, it is essential to clarify that classifications differ across countries and that specific terminology varies worldwide. For example, the reserve known as "frequency regulation" in North America is known as "frequency response" in the United Kingdom and as "primary control reserve" in Germany and most of mainland Europe. Table 9.1 provides an overview of utility-scale energy storage applications that can be commercially explored and potentially exploited.

In a brief overview, the five relevant power markets accessible to V2G/BESS exploration are: frequency regulation (FR), spinning reserves, transmission and distribution, voltage support, and energy arbitrage.

The FR market is the one which requires the shortest provider response times and is called upon to balance the system frequency (measured in cycles

Table 9.1 Possible markets for energy storage applications

Application	Capacity requirement	Classification	Discharge cycles per year
Peak PRICING ARBITRAGE	4–6 h	Bulk Storage	200–400
Generation Capacity	2–6 h	Bulk Storage	200–600
Transmission and distribution (T&D) Asset Capacity	2–4 h	Bulk Storage	201–600
Frequency Regulation	1–15 min	Ancillary/Power Services	1,000–20,000
Volt/VAR Support	1–15 min	Ancillary/Power Services	1,000–20,000
Renewable Ramping/ Smoothing	1–15 min	Ancillary/Power Services	500–10,000

Source: adapted from Eller and Gauntlett [10]

per second, or Hertz). It deviates from the nominal value of 60 Hz in North America (and 50 Hz throughout much of the rest of the world). Control signals in FR markets are often given in 2–5 s time-steps, with reaction times of a few minutes or less. These technological requirements are well beyond any BESS capabilities [11]. The authors state that because of the high market-clearing prices for the service, early work in V2G assumed that FR would be the preferred service for vehicle batteries. However, FR has one disadvantage: the high-energy throughput of a typical FR signal can result in significant battery degradation [12]. One must take into account that the EV owner should receive some extra remuneration to cover the excess degradation.

Spinning reserves are defined as "additional generation available to serve load in the case of an unplanned event, such as a power outage". As a result, it is a capacity service dispatched on an as-needed basis rather than regularly. On the other hand, spinning reserve providers are paid for having power available, not just when a dispatch response is issued. Because market values for spinning reserves are typically lower than those for FR, they have attracted much less research [11]. They may, however, be a better fit for markets where battery cycling degradation is a significant concern [13], since spinning reserves are used between 20 and 50 times a year [11]. In addition, this reserve can represent a more profitable service for V2G/BESS, because resources are generally called infrequently, and thus the price-to-power throughput ratio in the battery can be expected to be considerably higher than in the case of FR.

The simple approach of buying energy when it is cheap and selling it when it is expensive is referred to as "energy arbitrage" [11]. It can be carried out at the retail level using a retail time-of-day system or real-time energy prices. Retail customers in some jurisdictions have access to real-time retail pricing, while the large industry has access to it more frequently. Depending on the situation, it is possible to make money when the price difference is significant enough to overcome the round-trip energy losses and operational costs inherent in all forms of energy storage. It also should be noted that since the arbitrage does not involve direct contact with wholesale markets or grid administrators, retail energy arbitrage is treated as a V2H set-up. Thus, even smaller players with home-scale batteries can have access to this market.

9.2.3 Policy proposals for V2G

Given that the V2G scenario is still full of barriers, we discuss alternative regulatory approaches next, and explore how the elements need to be solved or clarified to incorporate V2G efficiently into a regulatory framework. We brought solutions from different regions where some testing was already done, particularly regarding how to integrate V2G and to ensure full access to the range of different electricity markets.

In Denmark, for example, the implementation of V2G faces two major economic obstacles: the high electricity price (of about 0.31 €/kWh) and increased taxes on electricity tariffs that are applied to both the charging and the discharging of an EV. As part of the high costs of the Danish electricity prices, the regular consumer will pay around 67% of the final electricity price in terms of taxes, including levies and VAT. Furthermore, since the nature of a V2G operation is to vary the battery level at all times, taxes on both sides of the operation make gross layer margins challenging to achieve due to the high operational costs. In sum, taxes and fees are responsible for almost 75% of the operating costs of V2G in Denmark.

The solution to this situation would be to separate the electricity used for the regular driving operation of an EV from the V2G business, where the charge/discharge cycles are used in the power market. Another temporary solution to incentivize V2G adoption could be to exempt V2G operation from taxes when storing and selling renewable energy, and by providing financial support similar to the existing one for renewable power generation (e.g., in the form of guaranteed feed-in tariffs). Currently, there are no incentives for the grid-friendly operation of home batteries together with renewable energy systems. As pointed out at the beginning of this section, it benefits particularly renewable energy usage, storing the electricity during peak supply periods for use during high demand periods, increasing both the value and the power mix of renewable energy. In the case of Denmark, since wind energy in that country is responsible for a large proportion of electricity generation, batteries could help to further increase the share of this source in the power mix.

9.3 OPERATIONAL PARAMETERS: MODELLING AND ENERGY MANAGEMENT ISSUES

Section 3.1 approaches the technical aspect of communication and aggregation structures between the EV owners and the aggregators. Section 3.2 addresses regulatory and legal concerns related to V2G, such as battery degradation and aggregation.

9.3.1 Aggregation and communication

Once the regulatory hurdles are overcome, we now must explore possible issues that originate from the two most important entities in this scenario: the EV owner and the utility industry.

Compared to a domestic battery that is typically used with a photovoltaic system, the EV battery is considerably more prominent; however, one single EV is not enough to provide the necessary energy for any of the markets. Usually, in many countries, 1 megawatt (MW) is the minimum required participation capacity for EVs [14]; this means that we need approximately

20 cars with an average battery size of 58.5 kWh in order to reach the minimum size demanded by the market. However, other markets may have a lower minimum requirement and can adapt their necessities accordingly.

Therefore, a third and important part and actor is the so-called "aggregator". The aggregator is the central actor who controls, coordinates and optimizes all the required operational activities, such as the communication with the distribution system operator (DSO), the transmission system operator (TSO), and the energy service providers [15].

In addition to that, an aggregator makes the connection between several EVs in order to form a single entity allowed to negotiate its energy in the market. This integration can be formulated within a virtual power plant (VPP), in which the E.V.s are controlled as a single distributed energy source [16]. Through aggregation, the aggregator can implement control and dispatch algorithms that provide the means to optimize various aspects of the battery storage systems and enhance the profitability of operations. The aggregator can also facilitate the bidirectional charging, be part of the initial investment, and look for a possible EV owner to join the market.

9.3.2 Technical challenges

Among the most major constraints of V2G deployments is the technological landscape necessary to arrange a smooth connection between numerous agents. Without a completely integrated technological system being present in the E.V., the charger, and other communication protocols, the successful achievement of V2G seems unlikely. In addition, both software and hardware components are expensive and complex, and in short supply at present.

In this sense, V2G operation as a technological innovation can be considered a part of the "big data" or the so-called "internet of things" (IoT). This complexity naturally brings challenges associated with modern applications, such as collecting, storing and managing a vast amount of data and securing it to maintain the privacy of its users to the extent required. Noel et al. [5] provides an example: If a V2G system has one million EVs connected and the system is collecting data in every second, the result is billions of binary points measured in an hour. Thus, for every million cars in an EV fleet, it is estimated that about a trillion kilobytes of information needs to be stored.

Regarding the security of this information, some consumers have expressed their concerns related to their personal data collection. Therefore, it is vital to regulate data collection and establish ideal conditions, so that EV owners can feel secure in sharing their data with energy companies and other service providers.

Due to innovative grid applications, such as V2G, numerous emerging control and information technologies are being integrated into the modern or "smart" power grid. However, in a worst-case scenario, smaller parts of the systems (e.g., smart vehicles) can be the entry door for an attack. The main concern is how far the attack can reach the main system and destabilize a larger region, leading to huge and immediate economic damage and probably pushing EV owners out of the business [17].

Another challenge surrounding the application of EVs and BESS into the power system is the development and implementation of the necessary communication protocols that handle the exchange of charging data and trade information between EVs and the electricity grid, and thus connecting the units and the aggregators. A good communication standard is intimately connected to good protocols of security and privacy mentioned before, since a properly designed communication can ensure both.

Communication protocols are currently standardized and proposed by different groups, such as the International Organisation for Standardization (ISO). For example, ISO 15118 is the international standard concerning "Road Vehicles – Vehicle to the grid communication interface." The main goal is to outline the digital communication protocol that an electric vehicle supply equipment (EVSE) and charging station should use to recharge the EV's high-voltage battery. ISO 15118 covers all charging-related use cases across the globe, including wired (AC and DC) and wireless charging applications and the pantographs used to charge larger vehicles such as buses [18]. This intelligent, two-way contact between the energy grid (operators and markets) and the vehicle (and driver) helps utilities to control electricity resources while also enabling vehicle owners to lower their cars' life cycle costs by selling power back to the grid [19].

Important to acknowledge is also that, despite the prestige of the ISO institution, the novelty and importance of the mentioned regulation, ISO 15118 is still the object of much discussion and controversy. Kester et al. [19] for example, criticize the unsatisfactory security requirements concerning confidentiality, authenticity, and privacy protection. It is also a target of criticism that only unilateral authentication (on the server side) is mandatory, while mutual authentication (the EV side) is not.

Table 9.2 presents a variety of different existing international norms. For example, SAEJ2847 and SAEJ2836 along with SAEJ1772 specify the communication requirement between an EV and the charging infrastructure. SAEJ2847 specifies the communication requirements, whereas SAEJ2836 defines the use cases and provides the testing infrastructure [20].With five different versions depending on the power, CHAdeMO is a DC charging standard for EVs, enabling seamless communication between the car and the charger. It has been developed by the CHAdeMO Association, which is also authorized to carry out the certification. CHAdeMO was proposed in 2010 as a global industry standard by an association of the same name formed by five major Japanese carmakers [21].

9.3.3 Battery degradation

The natural degradation of the battery and the accelerated degradation due to its increased use are the two most difficult challenges in making battery

Table 9.2 Different standardization norms, their emphasis and regions where applied

Norm	Focus	Description	Region
SAE J2847	E.V.–E.V.S.E.	A de-jure standard with three parts. Of these parts, part 1 (AC charging) and 3 (V.2.G.) are based on SEP 2.0, and part 2 (unidirectional DC charging) is based on ISO 15118 and PLC.	United States
CHAdeMO	E.V.–E.V.S.E.	A de-facto, by now globally recognized, DC Quick charge standard, which builds on ISO 61851 but then uses alternative hardware and communication protocols based on Controller Area Network (CAN) not PLC.	Japan/ Korea
GB/T 27930 (2015)	E.V.–E.V.S.E.	A de-jure standard. V.2.G. aspect primarily based on SAE J2847, but with CAN based communication protocol following the Original Equipment Manufacturer (OEM) practice of using ISO 11898.	China
ISO 61850 & ISO 61851	E.V.S.E.-Grid	A de-jure standard. Used for low level (automated) grid communication between and towards electricity substations. Focuses on voltage, frequency, and duty cycles	Globally
OCPP	E.V.S.E. Operator	A de-facto standard. Structures data streams between E.V.S.E.s and control servers	Western Europe
OCHP	E.V.S.E. operators	A de-facto standard. Structures data streams between E.V.S.E. operators and an international clearinghouse to allow for interoperability between different charging station operators.	Western-Europe
SEP 2.0	Smart grids	Developed by the ZigBee Alliance. General PLC based machine-to-machine communication protocol for energy appliances (incl. E.V.s) within a smart grid	United States

Source: Adapted from Kester et al. [19]

storage systems available [22, 23]. This issue can result in capacity loss over time, reducing EV range and affecting consumers' willingness to adopt such vehicles. Furthermore, using the battery to provide additional services, such as V2G system participation, may cause additional degradation and drive participants away from new business models. The economic feasibility of V2G systems is also affected by battery degradation. Accordingly, the payment offered to consumers must be high enough to compensate at least any additional deterioration or need for battery replacement [24].

It is important to explore the generic deterioration of the EV battery prior to exploring how it will be affected when offering additional V2G services.

Since the battery is the single most expensive component in an EV, it seems reasonable to assume that consumers will give more attention to this specific feature. Battery degradation is assumed to reach its end-of-life (EOL) when the battery's remaining capacity can provide only between 20% and 30% of the original storage capacity [25]. However, new studies have introduced up a new perspective, since personal preferences seem to be more relevant than technical limitations. For instance, the driving range in the United States rarely exceeds 100 km of daily travel, so in a normal scenario between 85 and 89% of drivers can be satisfied with EV charging at home using standard 120 V wall outlets. In addition to that, in their normal daily driving between 77 and 79% of the drivers will have over 60 km of buffer range for unexpected trips [26].

After the battery has reached the level where it is no longer considered as being sufficient for EV use, it will need to be replaced, possibly resulting in a substantial additional cost for the car owner. However, many studies have considered using depleted batteries in a second life as stationary batteries to provide grid services in order to offset the high costs of batteries [26].

The risk of accelerated electric vehicle battery degradation is commonly stated as one of the major concerns inhibiting the implementation of V2G technology.Still, nowadays the most popular cars support V2G technology, such as the Renault (Zoe) and the Nissan (Leaf) [27]. Egbue and Long [28] showed that even though EVs on the market currently have battery warranties of around 8–10 years, 90% of the respondents said they still have worries about possible failure or battery degradation.

Nevertheless, it is possible to make some assumptions and predict battery behaviour under normal conditions. Figure 9.2 shows average battery degradation of 100 EVs with different charger availability; although the capacity is not linear, we assume an average loss of 3% per year, so that after ten years, the degradation amounts to 31%, which is substantial [29].

While both calendar and cycling ageing play a role in battery degradation, the temperature has the greatest impact on both forms of ageing, causing high capacity loss and battery degradation [4, 30].

Given this scenario, Wang et al. [29] investigated the matter and carried out a comparison of EV battery degradation in two different scenarios: (i) driving and uncontrolled charging and (ii) driving and providing many V2G services (peak-load shaving, frequency regulation, and net-load shaping[1]). The results show that even in extreme cases, where the EV provides FR and peak-load shaving every day for ten years, no significant accelerated battery degradation was found compared to the uncontrolled charging scenarios. The battery degradation costs were estimated to be between $0.38 and $0.20 for the case of two hours of peak-load shaving and FR services.

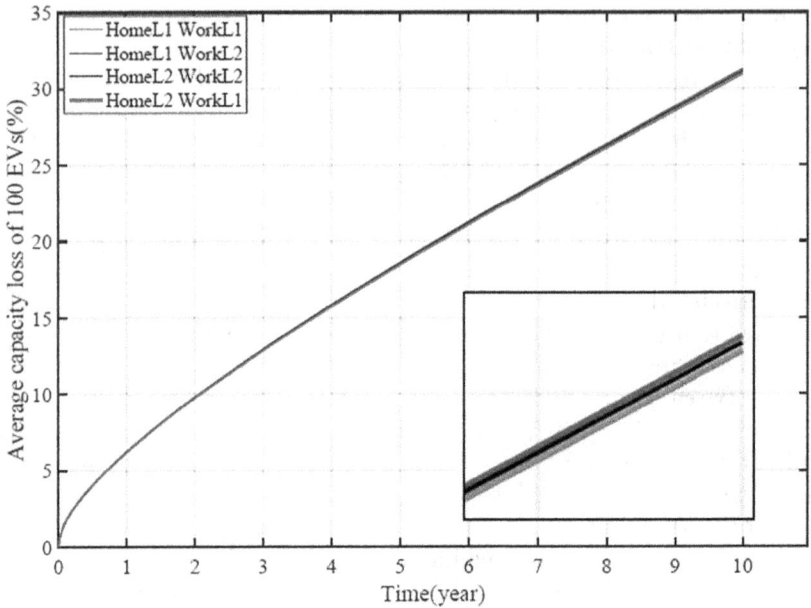

Figure 9.2 Average battery degradation of 100 EVs with different charger availability (Source: adapted from Wang et al. [29].

Calearo and Marinelli [31] analyzed how profitability is affected concerning the battery degradation when operating FR services. In Japan and
Denmark, when it comes to battery deterioration, in the case when frequency deviates from the norm, the losses are greater because the battery is
subject to more cycles over its lifetime. Nonetheless, when analyzing residential electricity prices, the battery degradation costs are about 1% of the
total costs, and 3–8% when considering industrial pricing, with both 55%
and 75% state of charge (SOC).

The complexity involving battery degradation becomes even higher
because other aspects not related to driving or V2G business may affect battery duration. These factors include temperature, weather conditions, and
the driving and charging habits of the EV owner.

9.3.4 Charging parameters

The stochastic and unpredictable nature of EVs, such as departure/arrival
times to charging stations, daily driven distances, battery sizes, and charger/
discharger types, are all influential parameters that can reduce power grid
reliability [32].

In addition, charging an EV also entails technical challenges concerning
energy efficiency, e.g., how much energy is lost during energy transportation
from the battery to the grid and vice versa. First, it is important to mention

that EVs already have a considerably high efficiency compared to their sustainable pairs, such as hydrogen; while the EV efficiency is around 70%–80%, hydrogen systems have an efficiency of only 40%–50% [4].

We can notice that the profitability of a V2G system will be highly affected by the technical efficiency of all components in this complex system. Therefore, Apostolaki-Iosifidou et al. [33] empirically measured and analyzed the power losses from an EV integrated into the electric system via measurements of different components, such as the building circuits, power feed components, and some sample electric vehicles. Under the conditions studied, the results show that one-way losses vary between 12% and 36%; moreover, they found that predominant losses were present in the power electronics used for AC-DC conversion. With two new design approaches, optimal sizing of charging stations, and a dispatch algorithm for grid services, the losses decreased by 7.0%–9.7%.

In summary, it is important to explore the generic deterioration of the EV battery prior to exploring how it will be affected when offering additional V2G services. For instance, Vonsien and Madlener [34] find that concerning battery ageing, smaller cycles cause a lower degradation to the battery compared to larger ones (normally associated with V2G activities). Another ageing-relevant parameter when dealing with charging is the c-rate – a measure of the rate at which a battery is discharged relative to its maximum capacity. If the speed of loading and discharging the battery is increased, higher throughput can be achieved and thus the economic efficiency of the battery can be increased temporarily. However, ageing effects occur faster when the manufacturer's specifications are exceeded, and hence the battery life is reduced, causing additional (battery-ageing) costs that reduce the achievable economic gains [34].

9.4 MAXIMIZING RES UTILIZATION AND INTERACTION WITH THE VIABILITY OF V2G/BESS

Further above, we defined and clarified the regulatory complexities of V2G and energy storage. The next challenge is to make the markets attractive and profitable to actors with stakes in energy storage. Therefore, this section addresses features that have an economic impact on the viability of V2G/BESS, such as financial policy support and the creation of new markets.

9.4.1 Value of stacking

As already discussed before, the introduction of VRE incentivized the participation of new actors in the market. Currently, this participation is still low, but it is assumed that energy storage will play a bigger role as the energy transition progresses, bringing flexibility and supporting the integration of high shares of variable renewable energy generation.

It is possible to bundle services in order to improve investments in flexible value chains for new technologies as an alternative. For example, aggregators can optimize value in the explicit flexibility case by offering many flexibility services to one or more market agents based on a single portfolio of cumulative flexibility from a group of prosumers [35].

The value of stacking is defined here as "the bundling of grid applications, which will boost the economics of distributed energy infrastructure by generating many value streams" [35]. One of the most important benefits of V2G/BESS is the opportunity for leveraging the same equipment, system, or process to deliver multiple benefits that maximize the financial gains. The economics of VRE infrastructure projects can be improved by stacking grid applications, increasing the return on investment, and shortening the payback time. Among these services may be demand response, ancillary services and self-resiliency (when the storage asset is used against unplanned power interruptions), etc.

9.4.2 Value creation by offering different grid services

As stated before, energy storage can generate much more value when different value stacked services are provided by the same device or by a fleet of vehicles [35]. Customers often benefit from the current behind-the-meter energy storage business model. However, those systems are currently installed for one of three purposes: (1) demand charge reduction; (2) backup power; or (3) increased solar self-consumption [36].

As a result, batteries are left unused or underused for more than half of the system's lifespan. In addition, calculating the profitability of home storage appliances, or the domestic use of an EV for household services, becomes highly dependent on the household profile, electricity prices, electricity self-consumption, and whether or not backup power is considered. On account of that, we examine the services offered to utilities or system operators, such as energy arbitrage, FR, spinning reserves, voltage support, and black-start. The description of these services is provided in Table 9.3.

Of these ancillary services, FR appears to be the most suitable for V2G operation [12]. This suitability occurs because frequency control does not need a large battery capacity and allows for shallow charge/discharge cycles rather than deep DoD, a measurement of battery capacity that is likely to shorten the battery's lifetime. A lithium-ion battery's number of cycles may be calculated as a function of DoD, implying that reducing DoD can avoid rapid battery deterioration. As a result, supplying EVs with frequency control services that have a lower DoD will have a reduced influence on battery-cycle lifespan [37–39].

Regarding FR, TSOs have to ensure a stable operation within their control area. To avoid this, the national TSOs use different types of frequency control reserve: frequency containment reserve (FCR), automatically activated frequency restoration reserve (aFRR), and manually activated frequency restoration reserve (mFRR). These reserve types differ according to their principle of activation, activation speed, and minimum lot sizes. These reserves compensate for sudden disturbances within seconds (FCR), five minutes (aFRR), or a quarter of an hour (mFRR).

FCR power has to be provided symmetrically in a positive and a negative direction, depending on the power frequency deviation from its nominal value. This means that electricity has to be injected into the grid (e.g., by ramping up generation units or discharging storage units) or that electricity has to be consumed. Hence, a unit providing 1 MW of FCR power must have a minimum rated capacity of 1.25 MW. By applying the degrees of freedom specified previously, the provider ensures adherence to the permissible range of SOC.

The call for tenders for FCR is "asymmetrical", meaning there is no separate call for tenders for positive FCR (additional power) and negative FCR (less power). The working day D-2 tender with a product period of one day, launched on 1 July 2019, has been replaced by a daily FCR tender with four-hour products from 1 July 2020 onwards [40]. In addition, a recent change of these requirements obliges BESS to provide FCR with prequalified power output for 15 minutes. The calculations are based on data series for the last 12 months regarding the FR and market data used in this work.

9.4.3 Parameters adopted

To increase the value of flexibility, market regulation still needs to adapt current rules and to enable the more efficient and more profitable operations of aggregators. The key regulation element must be able to stack value, which means aggregators must be able to offer multiple services to one or more balance-responsible parties from the same portfolio [35]. Examples include the following:

- in terms of time, allowing different services to be provided at different times. For example, in the morning, providing aFRR balancing service to the transmission system operator (TSO) and, in the afternoon, providing a congestion management service to the distribution system operator (DSO).
- in terms of pools, allowing entities to form a larger portfolio over a short period of time, or activating one asset or pool for one service and another asset or pool for a different service.
- in terms of double serving, by providing multiple services during the same period through stacking activation from one asset, pool, or portfolio. This form of value stacking can be divided into single energy transactions and double serving with multiple energy transactions.

Table 9.3 Services considered and definitions

Service	Definition
Frequency Regulation	The immediate and automatic response of power to a shift in locally sensed system frequency, from either the system or its components. To avoid frequency-level changes and grid instability, regulation is needed so that system-wide generation is perfectly matched with a system-level load on a moment-by-moment basis.
Spinning Reserve	Refers to generation capacity that is online and ready to serve load in an unplanned contingency event, such as a power outage. The non-spinning reserve is generation capacity that can respond to contingency events in a limited amount of time, usually less than ten minutes, but is not available immediately.
Voltage Support	Voltage control ensures that electricity flows reliably and continuously through the power grid. To ensure that both actual and reactive power output is balanced with demand, the voltage on the transmission and distribution system must be kept within a reasonable range.
Black-Start	In the case of grid outage, black-start generation assets are needed to put the regional grid back online by restoring service to larger power plants. Large power plants, in some cases, are capable of black-starting.

9.4.4 Business models

The V2G revenue is calculated using price signals from the German energy markets, a set driving profile, and an estimation of the battery degradation model to estimate degradation costs.

Since the target market for this work is Germany, and to make the profit calculations more realistic, the price signals are taken from the Regelleistung website [40]. The price signal for the year 2020 (1 January 2020 to 31 December 2020) is considered, and this price signal will be the input to the profit calculator.

The driving pattern of an EV varies between users and is highly unpredictable. Emergencies or unexpected plans can block the usage of EVs beyond the usual routine. However, we assume here the real-life situation that occurs on the contracts between the EV owners and energy companies, and the car owner commits to parking the car and connecting it to the grid every day at the agreed hours. The vehicle is used to drive from home to the office every morning, five days a week. For example, the EV is used at home between 18:00 and 07:00 hours; in the morning, it is handed over charged after it is has been driven to the office, parked between 08:00 and 17:00 hours, returned home for an hour, and parked at home between 18:00 and 07:00 hours when the cycle repeats. The EV is connected both to the office building and to the home charging point for the driving pattern considered.

Hence V2G can occur either from 08:00 hours to 17:00 hours at the office or from 18:00 hours to 07:00 hours at home. The maximum charging limit is 100% of the battery capacity, and the minimum remaining power at any point in the battery is 20%.

As mentioned before, since a single EV cannot meet the power requirements for the ancillary services market, an aggregator is believed to exist between the grid and the EV. The aggregator combines many EVs to meet FCR, aFRR, and mFRR power requirements. In total, there are seven price signals to be considered for the daily calculation: One price signal for the FCR and three each for aFRR and mFRR (3 different bidding slots of 4 hours each with positive and negative frequency control). The remaining price signals were not considered because they conflicted with the use of the EV for driving or charging.

Figure 9.3 depicts the daily expected behaviour of driving, charging, and market operations.

The revenue model is shown in Equation (9.1). Here, we define revenue as the income generated by the sale of V2G services. In this case, the revenue earned is the amount paid to the EV owner for the electric energy discharged (sold) to the grid, i.e.

$$R_{v2gpoll} = \sum_{n=1}^{N} * \left(\frac{\Delta SOC\_pool}{N} \right) * price\_signal(n) * \\ pool\_battery\_capacity * eff \tag{9.1}$$

where ΔSOC_pool denotes the percentage of the EV pool capacity discharged to the grid; N is the number of hours of discharge; Price_signal is the selling price of the electricity for the corresponding hour; Pool_battery_capacity represents the total capacity of the battery pool; and Eff is the the discharge efficiency, assumed here to be unity.

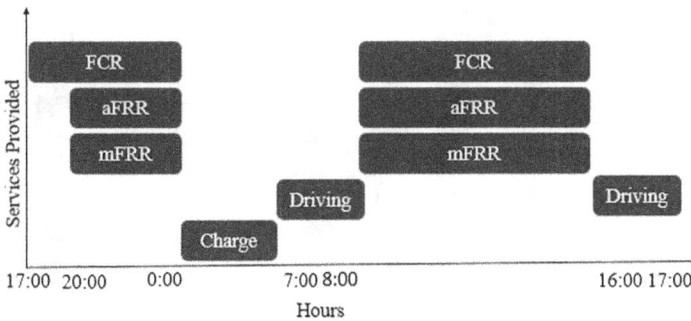

Figure 9.3 Services provided throughout the day.

9.4.5 Results

As mentioned previously, this chapter aims to create a systematic allocation to the appropriate market and product. The delivering and solving of many restriction problems and seeking the most suitable economic result is done through the stacking value, which depends on double serving with multiple energy transactions. We present the results where an aggregator could offer three different services (FCR, aFRR and mFRR) simultaneously. For all the cases, a minimum size of 1 MW is assumed.

Figure 9.4 depicts the results found by our model. Considering a V2G operation in different reserve types in the ancillary services market for 365 days – 12 hours per day, and a maximum revenue of €27,669 per year is estimated for the year of 2020, we can see the result is similar to previous studies conducted under similar conditions [12, 41, 42].

One can notice from Figure 9.4 that the average monthly revenue for 1 MW is around €2,000. However, in May and October, we see spikes in the price, causing a substantial increase in the achievable revenue. This increase is due to single participation in the mFRR market on 4 May for the first case. For the second case, prices were generally higher since the end of September, and, on many occasions, both aFRR and mFRR featured higher prices. As a result, while the average daily revenue was around €75 per day, revenues in some time slots in October reached about €400.

Concerning the source of revenue, FCR was responsible for around 60%, mFRR 28%, and aFRR 12%. Figure 9.5 shows the revenues per type of reserve in absolute terms.

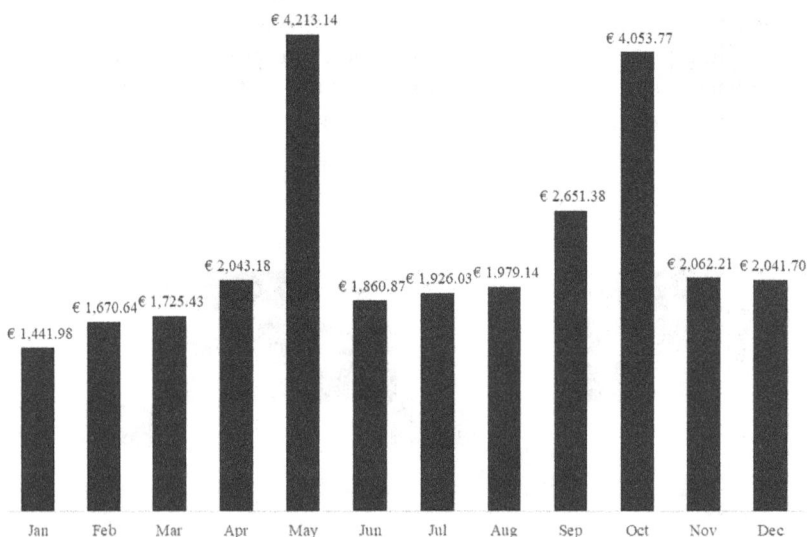

Figure 9.4 Monthly revenue from V2G operation, 2020.

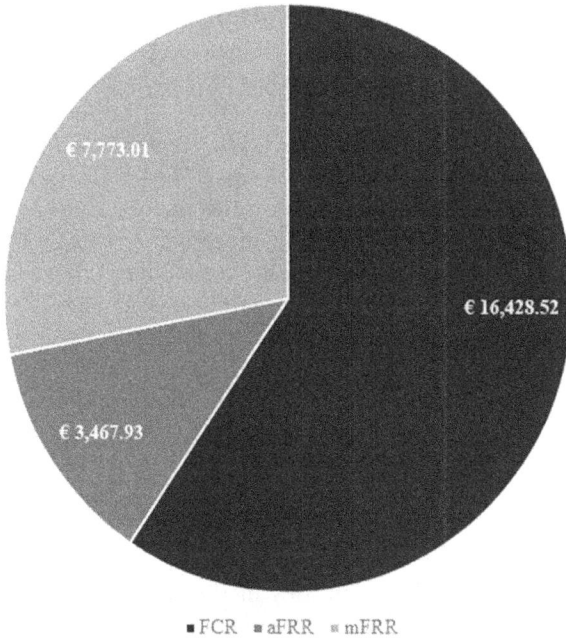

Figure 9.5 Revenue per type of reserve used for V2G operation, 2020.

Due to the limitation of 1 MW, it was impossible to participate in two different types of reserve markets simultaneously. That limited the revenue because, generally, aFRR and mFRR showed higher prices for the same hours (e.g. from 8:00 to 12:00 hours). However, in similar conditions, it can be inferred from the results that mFRR is more profitable than the aFRR service. This higher mFRR profitability is expected, as the average price of mFRR is greater than the aFRR service, although the hours they are activated do not differ substantially.

It is also important to mention that due to the different business models from V2G, it is still unclear as to how the relation between the aggregator and the EV owner is established. Hence, the remuneration costs to the car owner are still hard to calculate. The same can be said about the battery degradation and V2G. operation; in our model, we could operate 50% of the time due to driving and charging patterns. From the remaining time slots, we offered the service for two hours, keeping the battery at a normal level of degradation and without taking degradation into account [27].

Another variable to be considered is the size of the pool and the different cars that form it. The many existing differences in terms of battery size, the type of charger that the EV accepts, and the charging period could interfere with the pool size, making a significant difference in how the revenue is distributed and impacting the costs. Another major bottleneck for a V2G offer is expenditures for smart communication and optimization of the system.

9.5 CASE STUDIES OF PRACTICAL IMPLEMENTATIONS

In Section 9.4, we showed different economic results and how they relate to all the aspects discussed in the preceding sections – by discussing the past and current practical implementation of V2G, barriers found, and opportunities for improvement from a practical perspective. With the insights provided in the previous sections, governments and other actors interested in developing the V2G technology may want to incentivize societal action towards the acceptance and use of energy storage technologies. Implementing pilot projects is a good opportunity to test possible developments without causing hard disruption in the current framework. In this section, we describe selected pilot projects that have already been successfully implemented, and how their results can guide future decisions. These projects can provide useful insights and give experience to consumers; at the same time, new guidelines are prepared by governments.

9.5.1 The Parker V2G pilot project (Denmark)

Denmark is among the first countries in the world to have a fully commercialized V2G pilot project. Two initial projects (EDISON and Nikola) have already helped to provide a better understanding of the EV potential in balancing the Danish grid. With the Parker project, the objective is to validate that a series-produced EV as part of an operational vehicle fleet can support the power grid by becoming a vertically integrated resource, providing seamless support to the power grid both locally and system-wide. Additionally, the project seeks to ensure that barriers regarding the market, technology, and users are dealt with to pave the way for further commercialization, and, not least, to evaluate the specific EVs' capabilities to meet the needs of the grid [6].

The Parker project is the first to experimentally validate the ability to provide V2G services across various modern EV models. Validation is done using a test cycle for measuring the performance of EVs in response to a power request and testing some of the most demanding services today, such as FR. The project duration was from August 2016 to July 2018, and the partners include Nissan, the Mitsubishi Corporation, Frederiksberg Forsyning A/S, and DTU Electrical Engineering.

The most important barriers identified were the limited period over which a battery can operate, the inability to serve a non-balanced grid for long periods, the potential battery degradation, and the conclusion that cycles from the provision of the services will, to some extent, reduce the battery capacity. In addition, other barriers found were the challenging process of pre-qualification of aggregated EVs and the heavy tariffs and energy taxation in the country, especially regarding any two-way energy flows.

On the other hand, extensive tests by Parker have shown clearly that the technology works for the provision of additional grid services. Considering the price variations from different market reserves, the accounted annual revenue per car for 2017 was €1,711, and the 2018 revenue was €2,486. The higher price in the last year was because of the climate situation, with large periods of little to no wind, which resulted in a large uptick in the availability price.

The project discovered that electric vehicles with DC chargers and the CHAdeMO 2.0 protocol could provide valuable grid services by responding quickly, accurately, and precisely to bidirectional power requests.

9.5.2 Los Angeles Air Force Base

At the Los Angeles Air Force Base (AFB), an all-electric fleet pilot was introduced in 2013 with a limited fleet of 40 cars; each EV was also equipped with at least one compatible electric vehicle service equipment (charging station). The Los Angeles AFB fleet provided FR, a service in which a resource consumes or discharges electricity on command to match the electricity system demand to generation. This service contributes to grid stability by maintaining a consistent system frequency. During frequency control, "regulation up" and "regulation down" apply to discharging and absorbing energy, respectively. The Lawrence Berkeley National Laboratory developed a fleet scheduling, optimization, and control software framework jointly with Kisensum LLC to enable the vehicle fleet to compete in the California Independent System Operator's ancillary services markets [43].

This phase one pilot has been conducted to anticipate a larger Department of Defense non-tactical fleet electrification scheme. Therefore, the programme has two phases: the second phase includes some 500 medium-duty vehicles at six military bases, including one in Los Angeles, and the final fleet will be extended to about 1,500 vehicles.

For a total of 255 megawatt-hours of the regulation-up and 118 MW hours of regulation-down over 20 months, the demonstration successfully delivered FR to the California Independent Device Operator's market. In addition, the EV fleet performed well above the minimum power output needed, according to Lawrence Berkeley National Laboratory's implementation of the device operator's accuracy metric. Concerning the battery degradation, from May 2016 to August 2017, the team reported an average battery capacity loss of 5% to 10%. However, it was not possible to assess whether delivering ancillary services caused deterioration because there was so little difference in the usage of each electric vehicle in providing frequency control and travel that this effect could not be determined.

Concerning the revenues, even with the limited bid magnitudes, participation hours, and numbers of active EVs, the monthly revenue per vehicle was $124, and the total monthly revenue of the fleet was $5,097 on average. However, the authors point out how fees have a negative impact on the net profit, making it hard to scale the project on the national level.

9.6 CONCLUSIONS

This chapter has laid out several technical and economic challenges that BESS and V2G operations face both now and in the longer term. We showed that the current regulation is not yet adapted to new sources of flexibility, such as distributed energy storage. Although current regulation in some regions does not prevent new solutions from arising, it might prevent them from being less profitable and less often used. We also addressed the technical parameters that inherently affect the BESS/V2G, such as driving patterns, battery degradation and the many challenges involved in this operation – including optimization, aggregation of different actors, and data security.

We gathered this previous information and used Germany as a case study, and we calculated the revenues from a V2G operation in the country using the ancillary services market as the main operation source. The results show that the activity is feasible. However, many variables should be better explored, such as the remuneration of the EV owner, the ideal size of the EV pool, etc. Our final section presented real-life cases that have successfully implemented V2G operations in two different markets. Even though many barriers still prevent the fully commercial exploration of a V2G operation, many countries worldwide are already changing their regulation to embrace new forms of distributed energy storage, creating a very optimistic scenario for these operations.

NOTE

1 A service offered in California which aims at mitigating the fluctuation of net load and the influence of an increase in renewable energy penetration.

REFERENCES

[1] N. Patel, A. K. Bhoi, S. Padmanaban, and J. B. Holm-Nielsen, *Electric Vehicles Modern Technologies and Trends*. Springer, Singapore, 2021.

[2] IRENA, *Innovation landscape for a renewable-powered future: Solutions to integrate variable renewables*. International Renewable Energy Agency, Abu Dhabi, 2019.

[3] H. Lund and W. Kempton, "Integration of renewable energy into the transport and electricity sectors through V2G," *Energy Policy*, vol. 36, no. 9, pp. 3578–3587, 2008.

[4] L. Noel, G. Z. de Rubens, J. Kester, and B. K. Sovacool, "Beyond emissions and economics: Rethinking the co-benefits of electric vehicles (EVs) and vehicle-to-grid (V2G)," *Transport Policy*, vol. 71, pp. 130–137, 2018.

[5] L. Noel, G. Z. de Rubens, J. Kester, and B. K. Sovacool, "Navigating expert skepticism and consumer distrust: Rethinking the barriers to vehicle-to-grid (V2G) in the Nordic region," *Transport Policy*, vol. 76, pp. 67–77, 2019.

[6] Andersen Peter, Hashemi Seyedmostafa, Sousa Tiago, Meier Soerensen Thomas, Noel Lance, and Christensen Bjoern, "The parker project: Cross-brand service testing using V2G," *World Electric Vehicle Journal*, vol. 10, no. 4, 2019, doi: 10.3390/wevj10040066

[7] P. Bach Andersen, S. Hashemi, T. Sousa, T. Meier Soerensen, L. Noel, and B. Christensen, "The parker project: Cross-brand service testing using V2G," *World Electric Vehicle Journal*, vol. 10, no. 4, p. 66, 2019.

[8] G. Angenendt, S. Zurmühlen, H. Axelsen, and D. U. Sauer, "Comparison of different operation strategies for PV battery home storage systems including forecast-based operation strategies," *Applied Energy*, vol. 229, pp. 884–899, 2018.

[9] S. Englberger, H. Hesse, D. Kucevic, and A. Jossen, "A techno-economic analysis of vehicle-to-building: Battery degradation and efficiency analysis in the context of coordinated electric vehicle charging," *Energies*, vol. 12, no. 5, p. 955, 2019.

[10] A. Eller and D. Gauntlett, *Energy storage trends and opportunities in emerging markets*. Navigant Consulting Inc., Boulder, CO, USA, 2017.

[11] N. S. Pearre and H. Ribberink, "Review of research on V2X technologies, strategies, and operations," *Renewable and Sustainable Energy Reviews*, vol. 105, pp. 61–70, 2019.

[12] Z. Li, M. Chowdhury, P. Bhavsar, and Y. He, "Optimizing the performance of vehicle-to-grid (V2G) enabled battery electric vehicles through a smart charge scheduling model," *International Journal of Automotive Technology*, vol. 16, no. 5, pp. 827–837, 2015.

[13] Jonathan Mullan, David Harries, Thomas Bräunl, and Stephen Whitely, "The technical, economic and commercial viability of the vehicle-to-grid concept," *Energy Policy*, vol. 48, pp. 394–406, 2012, doi: 10.1016/j.enpol.2012.05.042

[14] ENTSO-E, "Consultation report FCR cooperation," 2017. Available: https://www.entsoe.eu/Documents/Consultations/20170601_FCR_Consultation_Report.pdf

[15] F. Mwasilu, J. J. Justo, E.-K. Kim, T. D. Do, and J.-W. Jung, "Electric vehicles and smart grid interaction: A review on vehicle to grid and renewable energy sources integration," *Renewable and Sustainable Energy Reviews*, vol. 34, pp. 501–516, 2014.

[16] M. Musio, P. Lombardi, and A. Damiano, "Vehicles to grid (V2G) concept applied to a virtual power plant structure: The XIX International Conference on Electrical Machines-ICEM 2010," pp. 1–6, 2010.

[17] Yingmeng Xiang, Lingfeng Wang, and Nian Liu, "Coordinated attacks on electric power systems in a cyber-physical environment," *Electric Power Systems Research*, vol. 149, pp. 156–168, 2017, doi: 10.1016/j.epsr.2017.04.023

[18] K. Bao, H. Valev, M. Wagner, and H. Schmeck, "A threat analysis of the vehicle-to-grid charging protocol ISO 15118," *Computer Science-Research and Development*, vol. 33, no. 1, pp. 3–12, 2018, doi: 10.1007/s00450-017-0342-y

[19] Johannes Kester, Lance Noel, Xiao Lin, Gerardo Zarazua de Rubens, and Benjamin K. Sovacool, "The coproduction of electric mobility: Selectivity, conformity and fragmentation in the sociotechnical acceptance of vehicle-to-grid (V2G) standards," *Journal of Cleaner Production*, vol. 207, pp. 400–410, 2019, doi: 10.1016/j.jclepro.2018.10.018

[20] H. S. Das, M. M. Rahman, S. Li, and C. W. Tan, "Electric vehicles standards, charging infrastructure, and impact on grid integration: A technological review," *Renewable and Sustainable Energy Reviews*, vol. 120, p. 109618, 2020.

[21] Mouli Gautham Ram Chandra, Kaptein Johan, Bauer Pavol, and Zeman Miro, "Implementation of dynamic charging and V2G using Chademo and CCS/ Combo DC charging standard," *2016 IEEE Transportation Electrification Conference and Expo (ITEC)*, pp. 1–6, 2016, doi: 10.1109/ITEC.2016. 7520271

[22] S. B. Peterson, J. Apt, and J. F. Whitacre, "Lithium-ion battery cell degradation resulting from realistic vehicle and vehicle-to-grid utilization," *Journal of Power Sources*, vol. 195, no. 8, pp. 2385–2392, 2010.

[23] A. Millner, "Modeling lithium-ion battery degradation in electric vehicles," in *2010 IEEE Conference on Innovative Technologies for an Efficient and Reliable Electricity Supply*, 2010, pp. 349–356.

[24] L. Noel, G. Zarazua de Rubens, J. Kester, and B. Sovacool, *Vehicle-to-Grid: A Sociotechnical Transition Beyond Electric Mobility*, 2019.

[25] C. Heymans, S. B. Walker, S. B. Young, and M. Fowler, "Economic analysis of second use electric vehicle batteries for residential energy storage and load-levelling," *Energy Policy*, vol. 71, pp. 22–30, 2014.

[26] Samveg Saxena, Jason MacDonald, and Scott Moura, "Charging ahead on the transition to electric vehicles with standard 120V wall outlets," *Applied Energy*, vol. 157, pp. 720–728, 2015, doi: 10.1016/j.apenergy.2015.05.005

[27] Nissan, *Vehicle to grid: Residential V2G*. [Online]. Available: https://www.nissan.co.uk/range/electric-cars/v2g.html#

[28] Ona Egbue and Suzanna Long, "Barriers to widespread adoption of electric vehicles: An analysis of consumer attitudes and perceptions," *Energy Policy*, vol. 48, pp. 717–729, 2012, doi: 10.1016/j.enpol.2012.06.009

[29] Dai Wang, Jonathan Coignard, Teng Zeng, Cong Zhang, and Samveg Saxena, "Quantifying electric vehicle battery degradation from driving vs. vehicle-to-grid services," *Journal of Power Sources*, vol. 332, pp. 193–203, 2016, doi: 10.1016/j.jpowsour.2016.09.116

[30] J. Jaguemont, L. Boulon, and Y. Dubé, "A comprehensive review of lithium-ion batteries used in hybrid and electric vehicles at cold temperatures," *Applied Energy*, vol. 164, pp. 99–114, 2016.

[31] L. Calearo and M. Marinelli, "Profitability of frequency regulation by electric vehicles in Denmark and japan considering battery degradation costs," *World Electric Vehicle Journal*, vol. 11, no. 3, 2020, doi: 10.3390/wevj11030048

[32] Bijan Bibak and Hatice Tekiner-Moğulkoç, "A comprehensive analysis of Vehicle to Grid (V2G) systems and scholarly literature on the application of such systems," *Renewable Energy Focus*, vol. 36, pp. 1–20, 2021, doi: 10.1016/j.ref.2020.10.001

[33] Elpiniki Apostolaki-Iosifidou, Paul Codani, and Willett Kempton, "Measurement of power loss during electric vehicle charging and discharging," *Energy*, vol. 127, pp. 730–742, 2017, doi: 10.1016/j.energy.2017.03.015

[34] S. Vonsien and R. Madlener, "Li-ion battery storage in private households with PV systems: Analyzing the economic impacts of battery aging and pooling," *Journal of Energy Storage*, vol. 29, p. 101407, 2020.

[35] E. Klaassen, M. van der Laan, H. de Heer, A. van der Veen, and W. van den Reek, *USEF Whitepaper-Flexibility Value Stacking. Recommended Processes,*

Rules and Interactions to Enable Value Stacking for Portfolios of Flexible Demand-Side Resources. Available online www.usef.energy (Accessed on 29 March 2021).

[36] G. Fitzgerald, J. Mandel, J. Morris, and H. Touati, "The economics of battery energy storage: How multi-use, customer-sited batteries deliver the most services and value to customers and the grid," *Rocky Mountains Institute*, p. 6, 2015.

[37] W. Kempton et al., "A test of vehicle-to-grid (V2G) for energy storage and frequency regulation in the PJM system," *Results from an Industry-University Research Partnership*, vol. 32, 2008.

[38] A. de Los R'ios, J. Goentzel, K. E. Nordstrom, and C. W. Siegert, "Economic analysis of vehicle-to-grid (V2G)-enabled fleets participating in the regulation service market," in *2012 IEEE PES Innovative Smart Grid Technologies (ISGT)*, 2012, pp. 1–8.

[39] A. N. Brooks, *Vehicle-to-grid demonstration project: Grid regulation ancillary service with a battery electric vehicle.* California Environmental Protection Agency, San Dimas, CA, 2002.

[40] Regelleistung, *Internetplattform zur Vergabe von Regelleistung.* 2021. [Online]. Available: https://www.regelleistung.net/ext/ (Accessed on 14 February 2021.

[41] P. P. Malya, "Economic feasibility analysis of vehicle-to-grid service from an EV German electricity market," 2020. Available: https://elib.uni-stuttgart.de/bitstream/11682/10977/1/PrasadMalya_MT_print.pdf

[42] J. Jargstorf and M. Wickert, "Offer of secondary reserve with a pool of electric vehicles on the German market," *Energy Policy*, vol. 62, pp. 185–195, 2013.

[43] D. Black, J. MacDonald, N. DeForest, and C. Gehbauer, "Los Angeles air force base vehicle-to-grid demonstration: Final project report, California" 2020.

Chapter 10

A meta-heuristic-based optimal placement of distributed generation sources integrated with electric vehicle parking lot in distribution network

Mohd Bilal and M. Rizwan
Delhi Technological University, Delhi, India

CONTENTS

10.1 INTRODUCTION

The widespread use of traditional vehicles causes temperature rises as well as significant carbon dioxide emissions, both of which have a detrimental influence on the surrounding environment. As a result of non-uniform weather and climate changes induced by global warming, it is becoming more difficult to sustain the earth's natural system. To counteract the environmental

DOI: 10.1201/9781003311195-10

repercussions of traditional forms of transportation, an electric mode of transportation, i.e., battery-powered mobility, is required [1]. Electric vehicles (EVs) have several benefits, including the capacity to save money on carbon fuels and reduce pollution. There are several advantages for introducing and expanding EVs. Battery-powered vehicles are becoming increasingly popular as a method of decreasing pollution worldwide. As a result of this issue, EVs currently account for 28.8 percent of the vehicle in Norway, 6.5 percent in the Netherlands, and 1.5 percent in China. In addition, a number of countries want to adopt EVs as a mode of transportation in the future. According to current projections, over 35 million EVs will be on the road worldwide by the end of 2022. Aside from being environmentally benign, charging EVs has the potential to significantly improve the dependability of the electrical power system. As a result of the increasing system demand caused by EV charging, substation reserve capacity and feeder load transfer capacities are being reduced. The ability to transfer loads is particularly critical during system upgrades that necessitate the deployment of additional feeds. This has a significant impact on the system's overall reliability [2]. Furthermore, charging EVs with traditional power sources falls short of the aim of expanding EV adoption. The utilization of solar/wind energy to charge the battery of an EVs increases the benefits of electric automobiles.

Initially, EVs are connected to the power grid to recharge their batteries. New smart grid technologies offer distinct choices for energy transfer to the grid, known in this context as the vehicle-to-grid (V2G) mode. EVs that are connected to the electrical grid can be employed as energy storage devices [3].

Distribution system operators (DSOs) are responsible, according to European Union power laws, for providing amenities to linked consumers and distributed generation (DGs), as well as conducting network upgrades for electric vehicle parking lots (EVPLs) [4]. DSOs have no influence on charging locations or times. The unpredictable nature of EV charging makes system operation difficult to maintain grid stability. Unexpected conclusions were discovered in recent research on the influence of EVs' integration into power generation systems [5]. The impact of electric car charging on the power generation system, as well as the reduction of CO_2 emissions, are being discussed in [6]. The influence of EVs on emissions from the energy generating and transportation sectors is investigated using a hypothetical scenario.

In order to estimate the impact of EVs on the electrical network, EVs scenarios, and charge management methodologies are explored [7, 8] examines the effect of charging operations on the system's load profile. Various degrees of EV adoption are being studied to understand in order to the influence of EVs on system power losses [9]. The authors of [10] examined the effect of electric car charging on distribution transformer aging in rooftop solar photovoltaic (PV) installations. The installation of EVPLs has a negative impact on the distribution system, resulting in higher-than-necessary voltage and power losses. As a result, the incorporation of DGs has been promoted as a potential method for mitigating the charging implications of

EVs. Meta-heuristic approaches are frequently used to solve the EVPL and DG allocation problems in distribution networks. When compared to other methods, these are easier to implement, take less time to complete, have fewer variables to struggle with, and can provide significant outcomes. The suggested hybrid approach addresses these problems while being easy to implement and achieving high levels of convergent performance. As a result, this chapter will show how to distribute DGs in the presence of EVPL in a radial distribution network utilizing HS-TLBO, a hybrid intelligent approach. Hybrid algorithms utilize the best features of both strategies to achieve results that outperform either strategy on its own.

10.2 MATHEMATICAL DESIGN OF THE PROBLEM

In general, load requirement varies with time on the distribution end of the grid network; however, with changing load, the optimal location and size of EVPL and DG are not acceptable. As a result, in order to solve the problem of optimal DG and EVPL allocation planning, the following assumptions are used [11]:

(1) Radial distribution systems are naturally balanced.
(2) EVPL with constant capacity are installed.
(3) DGs with a power factor of one are employed.
(4) The output of the DGs is not time-varying.

Since DGs do not alter the bus voltage, they are treated as negative loads in this chapter. The best DG position and size should be achieved without violating the technical restriction, which must be evaluated using load flow analysis at each iteration. The primary objective functions that are explored include system losses and the enhancement of the voltage profile.

10.2.1 A Direct Approach Method for Load Flow

Most of the distribution networks are radial, with a low reactance-to-resistance ratio. In this chapter, power flow is accomplished using a direct search-based load flow approach [12].

The load flow technique used in this chapter is built around two matrices, which are named as the bus injection-to-branch current (BIBC) matrix and the branch current-to-bus voltage (BCBV) matrix. To determine load flow solutions, simply multiply these two matrices together. Figure 10.1. shows a simple 6-bus distribution network, with $P_1 - P_5$ representing branch currents and $Z_{12} - Z_{56}$ representing branch impedance.

It is possible to calculate the current injected at the ith node as follows:

$$I(i) = \left(\frac{P(i) + jQ(i)}{|V(i)|} \right)^{*}$$

(10.1)

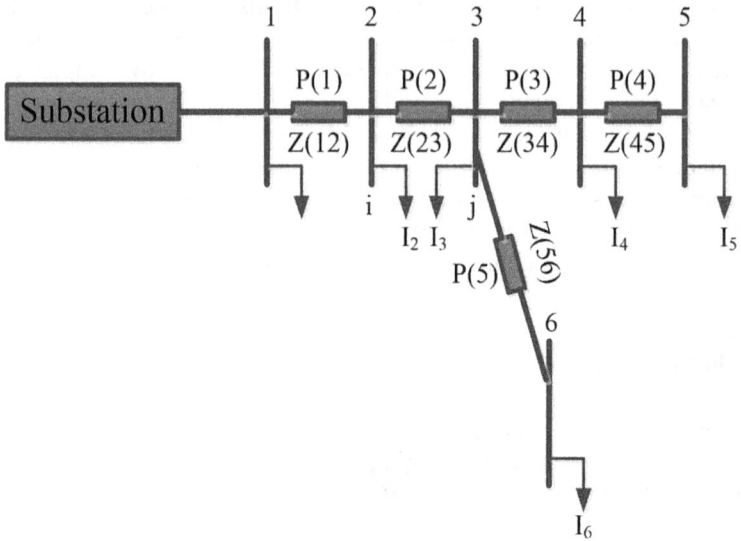

Figure 10.1 Typical diagram of a 6-node electrical network.

where $I(i)$ denotes the current, $P(i)$ and $Q(i)$ represent the active and reactive power injected at the ith node, respectively, and $V(i)$ indicates the voltage at i^{th} bus, respectively.

Equation (10.1) can be further fragmented into its real and imaginary parts.

$$Real\big(I(i)\big) = \frac{P(i)\cos\theta(i) + Q(i)\sin\theta(i)}{|V(i)|} \tag{10.2}$$

$$Imag\big(I(i)\big) = \frac{P(i)\sin\theta(i) - Q(i)\cos\theta(i)}{|V(i)|} \tag{10.3}$$

where $\theta(i)$ specifies the voltage angle at the ith bus.

B stands for branch current matrix, as shown in Figure 10.1, and matrix P is designed using Kirchhoff's current law, using Equation (10.3).

$$\begin{bmatrix} P(1) \\ P(2) \\ P(3) \\ P(4) \\ P(5) \end{bmatrix} = \begin{bmatrix} 1 & 1 & 1 & 1 & 1 \\ 0 & 1 & 1 & 1 & 1 \\ 0 & 0 & 1 & 1 & 0 \\ 0 & 0 & 0 & 1 & 0 \\ 0 & 0 & 0 & 0 & 1 \end{bmatrix} \begin{bmatrix} I(1) \\ I(2) \\ I(3) \\ I(4) \\ I(5) \end{bmatrix} \tag{10.4}$$

$$[P] = [BIBC][I] \tag{10.5}$$

The voltage difference at each bus in comparison to the reference bus are calculated using Kirchhoff's voltage law as follows:

$$
\begin{bmatrix} E(1) \\ E(1) \\ E(1) \\ E(1) \\ E(1) \end{bmatrix} - \begin{bmatrix} E(2) \\ E(3) \\ E(4) \\ E(5) \\ E(6) \end{bmatrix} = \begin{bmatrix} Z(12) & 0 & 0 & 0 & 0 \\ Z(12) & Z(23) & 0 & 0 & 0 \\ Z(12) & Z(23) & Z(34) & 0 & 0 \\ Z(12) & Z(23) & Z(34) & Z(45) & 0 \\ Z(12) & Z(23) & 0 & 0 & Z(36) \end{bmatrix} \begin{bmatrix} P(1) \\ P(2) \\ P(3) \\ P(4) \\ P(5) \end{bmatrix}
\tag{10.6}
$$

$$[\Delta E] = [BCBV][P] \tag{10.7}$$

Incorporating the matrix P value from Equation (10.4) into Equation (10.6), we get:

$$
\begin{bmatrix} E(1) \\ E(1) \\ E(1) \\ E(1) \\ E(1) \end{bmatrix} - \begin{bmatrix} E(2) \\ E(3) \\ E(4) \\ E(5) \\ E(6) \end{bmatrix} =
$$

$$
\begin{bmatrix} Z(12) & 0 & 0 & 0 & 0 \\ Z(12) & Z(23) & 0 & 0 & 0 \\ Z(12) & Z(23) & Z(34) & 0 & 0 \\ Z(12) & Z(23) & Z(34) & Z(45) & 0 \\ Z(12) & Z(23) & 0 & 0 & Z(36) \end{bmatrix} * \begin{bmatrix} 1 & 1 & 1 & 1 & 1 \\ 0 & 1 & 1 & 1 & 1 \\ 0 & 0 & 1 & 1 & 0 \\ 0 & 0 & 0 & 1 & 0 \\ 0 & 0 & 0 & 0 & 1 \end{bmatrix} * \begin{bmatrix} I(1) \\ I(2) \\ I(3) \\ I(4) \\ I(5) \end{bmatrix}
\tag{10.8}
$$

$$[\Delta E] = [BCBV][BIBC][I] \tag{10.9}$$

$$[\Delta E] = [DLF][I]$$

DLF is an acronym for distribution load flow, and it is used to calculate voltages at each bus in comparison to the reference bus. DLF is calculated using the two steps as given below.

Step 1:   Determine the branch impedance vector Z_b using the line data.

Step 2: Make a diagonal matrix with the branch impedance vector by setting all lower and upper values to zero except for the main diagonal positions, then multiply the generated matrix by I to get delta E.

10.2.2 Objective Functions

The main purpose of this chapter is to find the best nodes in the radial distribution system for EVPL and DG placement in order to reduce network active power losses, and monitor the voltage profile within required limits while taking into account all constraints. The energy required to charge EVs is provided by EVPL. The capacity of an EV battery is measured in kWh and ampere-hours (Ah). When charging an EV, the EVPL is designed to provide only true current [13]. When EVPL is installed on any distribution network bus, it increases only real power. As a result, EVPL should be installed at the node (bus) with the least amount of branch current. In this regard, Figure 10.2 depicts a distribution network section with EVPL on the $(k + 1)^{th}$ bus and already coupled load on the same bus drawing energy from the electrical system.

$$\text{minimize}\left(F_1(x), F_2(x) \right) x \in © \tag{10.10}$$

$$\text{subjected to } g_m(x) = 0\, m = 1, 2, ..., z \tag{10.11}$$

$$h_n(x) = 0\, n = 1, 2, ..., s\, x_{lower} \leq x \leq x_{upper} \tag{10.12}$$

where $g_m(x)$ denotes the equality constraints whereas $h_n(x)$ denotes the inequality constraint, the values z and s represent the amount of equality and inequality constraints, respectively, x_{lower} indicates the lower bound while x_{upper} denotes the upper bound of variables and © represents the search space of variable.

10.2.2.1 Active power loss

In an electric network, the distribution system is typically where the most power is lost, which has an impact on annual sales. As a result, active

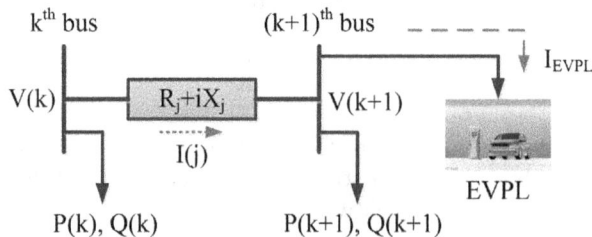

Figure 10.2 EVCS is connected at the bus of distribution system.

power loss minimization is the primary concern when allocating EVPL and DGs in radial distribution networks. To determine the active power loss or base case power loss, a distribution network load flow analysis is performed. This chapter conducts a load flow analysis using a direct approach [12]. The active power loss after the load flow is calculated using Equation (10.12), which is dependent on the amount of current drawn/injected into the bus [14].

$$\text{Active power loss} = \sum_{i=1}^{N_{br}} |P(i)|^2 * R(i) = Transpose(R) * |BIBC * I|^2 \quad (10.13)$$

where $P(i)$ represents the flow of current in the ith line, $R(i)$ represents the resistance of the ith line, and R represents the branch resistance matrix.

Active power loss increases when EVPL is deployed at any node in the distribution system because it acts as a high load. As a result, the goal is to select the bus with the smallest increase in active power loss. The distribution system's additional losses can be balanced by optimally ordering the DGs. The main purpose of DG units is to feed actual and reactive power into the system for the compensation of system losses induced by EVPL placement.

When I is divided into its actual and imaginary components, Equation (10.13) can be given as follows:

$$\text{Active power loss} = Transpose(R) * \left[\left(BIBC * actual(I) \right)^2 + \left(BIBC * Img(I) \right)^2 \right] \quad (10.14)$$

By applying the actual part of current from Equation (10.2) and imaginary parts of current from Equations (10.3) to (10.14), the overall real power loss can be calculated as follows:

$$\begin{aligned} \text{minimize } F_1 = Transpose(R) * \left(BIBC * \frac{P\sin\theta + Q\cos\theta}{|V|} \right)^2 \\ + Transpose(R) * \left(BIBC * \frac{P\cos\theta + Q\sin\theta}{|V|} \right)^2 \end{aligned} \quad (10.15)$$

10.2.2.2 Voltage Deviation Index (VDI)

The voltage deviation index is used to assess the voltage quality of the bus. Hence, bus VDI must be reduced in order to generate a more regulated bus voltage levels over the distribution network. In the proposed optimal EVPL and DG allocation, bus VDI is used as an objective function and is expressed as [15]:

$$\min F_2 = \sum_{k=1}^{N_{bus}} \left(V_k - V_{ref}\right)^2 \qquad (10.16)$$

The voltage magnitude on each bus must be between the minimum (0.95 p.u.) and maximum (1.05 p.u.).

10.2.3 Constraints

Below are the constraints that have been taken into account when allocating EVPL and DGs in the distribution network.

(a) Power balance

The active and reactive power demand, as well as the additional CS load capacity, must all equal the active and reactive power delivered by the electric substation and DG.

$$P^{substation} + \sum_{k=1}^{N_{bus}} P^{DG}(k) = \sum_{j=1}^{N_{br}} P_{loss}^j(k,k+1) + \sum_{k=1}^{N_{bus}} P_{D,k} + P_{EVCS}^k \qquad (10.17)$$

$$Q^{substation} + \sum_{k=1}^{N_{bus}} Q^{DG}(k) = \sum_{j=1}^{N_{br}} Q_{loss}^j(k,k+1) + \sum_{k=1}^{N_{bus}} Q_{D,k} \qquad (10.18)$$

where $P^{substation}$ is the real power and *and* $Q^{substation}$ is the reactive power supplied by the electric substation, respectively, $P_{D,k}$ is the active power demand and $Q_{D,k}$ is the reactive power demand at the kth bus, $P^{DG}(k)$ *and* $Q^{DG}(k)$ are the total real and reactive power injected by DGs at the kth bus, P_{loss}^j represents the actual power loss and Q_{loss}^j indicates the reactive power loss in the jth branch, P_{EVCS}^k is the EVPL load at the kth bus and N_{br} denotes the number of lines whereas N_{bus} represents the bus's number in the grid network.

(b) Voltage limit constraint

The voltage magnitude on each bus ranges between 0.95 p.u and 1.05 p.u.

$$V_{min} \le V_k \le V_{max} \, k = 1,2,....N_{bus} \qquad (10.19)$$

(c) Transmission line constraint

The actual flow of current in each line of network must not go beyond the maximum value of line current.

$$I_j \le I_j^{max} \, j = 1,2,3....N_{br} \qquad (10.20)$$

where I_j denotes the actual current flowing in the jth line and I_j^{max} denotes the maximum line current limit.

(d) Active and reactive power injected by DG

DGs must inject a certain amount of active and reactive power which is expressed as follows:

$$P_{DG_k}^{min} \leq P_{DG_k} \leq P_{DG_k}^{max} \tag{10.21}$$

$$Q_{DG_k}^{min} \leq Q_{DG_k} \leq Q_{DG_k}^{max} \tag{10.22}$$

$P_{DG_k}^{min}$ and $P_{DG_k}^{max}$ are the minimum and maximum active power limits of the kth DG respectively and $Q_{DG_k}^{min}$ and $Q_{DG_k}^{max}$ are the minimum and maximum active power limits of the kth DG.

(e) DG unit's penetration

$$\sum_{k=1}^{N_{DG}} P_{DG_k} \leq \%J \times \sum_{k=1}^{N_{bus}} P_{L_k} \tag{10.23}$$

where J denotes the maximum penetration of DG units in the distribution system and N_{DG} denotes the number of DGs installed in the system.

10.3 TEACHING-LEARNING-BASED OPTIMIZATION ALGORITHM (TLBO)

Rao et al. were the first to propose the TLBO algorithm in 2011. The TLBO algorithm is inspired by nature and is divided into two stages: teaching and learning. During the 'teaching phase', the best learner, i.e., the teacher, shares his understanding with the leftover learners in order to improve their understanding. Following that, each learner interacts with the other students in order to broaden his or her own knowledge during the 'learning phase'. These two phases are repeated until the TLBO finds the best global solution available.

10.3.1 Teaching Phase

The teacher continues to do his best to convey his expertise to the remaining students. The knowledge of the teacher can be used to enhance the remaining students/variables. This can be demonstrated using the example of achieving the goal of taming the understanding of students in a certain class in a definite subject. The highest-scoring student is designated as the teacher.

The teacher's job is to use his own scores to increase the old scores of other students to new scores that are closer to the mean scores of that subject/ variable. As a result, a random method is used to enhance the remaining students' grades in the class. New marks are generated for each individual by:

$$X_{new,i}^{HMS} = X_{old,i}^{HMS} + rand * \left(X_{teacher} - T_F M_i \right) \qquad (10.24)$$

where T_F stands for 'Teaching factor', which is either 1 or 2 [16]. If the fitness (marks) of the newly created solution vector (X_{new}) is better than the previous solution vector, it is approved. If the fitness (marks) of the newly created solution vector (X_{new}) is better than the previous solution vector, it is approved. The old and new variables in Equation (10.24) are $\left(X_{old,i}^{HMS} \right)$ and $\left(X_{new,i}^{HMS} \right)$, respectively, whereas 'rand' is a number generated at random between 0 and 1.

10.3.2 Learning Phase

Learners' understanding could be improved even further through their own efforts. Students in the class collaborate with one another at random, which improves their understanding of a subject. Equations (10.25) and (10.26) can be used to explain the learner phase mathematically:

$$X_{new,i}^{HMS} = X_{old,i}^{HMS} + rand * \left(X_j - X_k \right) if \ F\left(X_j \right) < F\left(X_k \right) \qquad (10.25)$$

$$X_{new,i}^{HMS} = X_{old,i}^{HMS} + rand * \left(X_k - X_j \right) if \ F\left(X_j \right) > F\left(X_k \right) \qquad (10.26)$$

F(X) represents a learner's fitness/knowledge (marks) in a given subject. A set of $X_{new,i}^{HMS}$ is accepted if the fitness corresponding to it is better than the fitness corresponding to the set of $X_{old,i}^{HMS}$; else, it is ignored.

An iteration is made up of these two phases, namely the teaching and learning phases. The TLBO algorithm ensures the global solution/best knowledge after many iterations of information exchange and sharing. Although, in comparison to other population-based optimization strategies, this methodology is almost parameter-independent; for multimodal problems, early convergence may occur due to strong local maxima/minima trappings.

10.4 HARMONY SEARCH ALGORITHM (HSA)

HSA imitates the improvisational skills of musicians. Just as the value of an objective function improves iteration by iteration, the sounds for greater aesthetic estimation can be improved via more and more practice. The HSA can be explained with the help of the following steps:

Step 1: **Initialization of the harmony memory (HM) and parameters of algorithm**

To begin, an HM matrix is created, with every row representing a collection of decision parameters $\left(X_{new,i}^{HMS}\right)$ for optimizing the objective function $F\left(X_{new,i}^{HMS}\right)$. Equation (10.27) is used to calculate each decision variable $\left(X_{new,i}^{HMS}\right)$.

$$X_{new,i}^{HMS} = X_{i,\min} + rand * \left(X_{i,\max} - X_{i,\min}\right) i = 1,2,3....N \qquad (10.27)$$

Where $X_{i,\,min}$ denotes the minimum marks and $X_{i,\,max}$ represents maximum marks.

HMS stands for the number of variables in harmony memory and is known as harmony memory size, HMCR denotes the likelihood of a new value for decision variables in harmony memory and is known as harmony memory consideration rate. The pitch adjustment rate (PAR) expresses the possibility of changing decision variables to adjacent values within a given range of possible values, and NI denotes the termination criteria.

$$HM = \begin{bmatrix} X_1^1 & X_2^1 & X_3^1 & ..... & X_N^1 \\ X_1^2 & X_2^2 & X_3^2 & ..... & X_N^2 \\ ..... & ..... & ..... & ..... & ..... \\ X_1^{HMS-1} & X_2^{HMS-1} & X_3^{HMS-1} & ..... & X_N^{HMS-1} \\ X_1^{HMS} & X_2^{HMS} & X_3^{HMS} & ..... & X_N^{HMS} \end{bmatrix} \qquad (10.28)$$

Step 2: **Improvisation of new harmony**

A set of new variables is formed on the basis of three essential factors: PAR, HMCR and random selection. Improvisation is the process of creating an original HM decision parameter. HMCR has a probability ranging from 0 to 1. It gives you the option of selecting a variable value from the harmony memory, whereas (1-HMCR) is the chance of selecting a variable from a list of alternative values, as illustrated in the following manner:

if rand < HMCR

$$X_{new,i}^{HMS} \leftarrow X_{old,i}^{HMS} \varepsilon \left\{ X_i^1 \ X_i^2 \ ......X_i^{HMS} \right\} \qquad (10.29)$$

else

$$X_{new,i}^{HMS} = X_{i,\min} + rand * \left(X_{i,\max} - X_{i,\min}\right) \ i = 1,2,3....N \quad (10.30)$$

end

HMCR determines whether or not to change the pitch of each variable. Following HMCR's consideration of the decision variable, PAR determines the pitch adjustment as follows:

if rand < PAR

$$X_{new,i}^{HMS} \leftarrow X_{old,i}^{HMS} + rand * BW \qquad (10.31)$$

else

$$X_{new,i}^{HMS} \leftarrow X_{old,i}^{HMS} - rand * BW \qquad (10.32)$$

end

Where BW is called as the bandwidth. The value of BW lies on the range of 0 and 1.

Step 3:   Updation of harmony memory
If the original set of decision variables outperforms the worst set of decision variables in HM in terms of objective function $F\left(X_{new}^{HMS}\right)$ values, the worst set is replaced.

$$X_{new}^{HMS} = \left\{X_1^{HMS}\ X_2^{HMS} \dots\dots\dots\dots X_N^{HMS}\right\} \qquad (10.33)$$

Step 4:   Termination criteria
The HSA repeats steps 1 to 3 until the maximum number of improvisation (NI) specified during the initialization phase is reached.

Before TLBO and HS are hybridized, numerous improvements to the algorithms are proposed so that the individual algorithms have improved exploration and exploitation capabilities.

10.5 MODIFICATIONS IN THE TLBO AND HS ALGORITHMS

10.5.1 Modifications in the TLBO algorithm

Some changes are made to improve the TLBO's convergence and avoid strong local minimum and maximum trapping.

10.5.2 Modifications in the teaching phase

An original collection of variables is predicated on the average marks of particular subject and the teacher's knowledge. The proposed method takes into account the solution vector with the lowest fitness rather than the mean M_j. This certifies that the weakest student in the class improves his or her knowledge, resulting in better information transfer to other students in the class. As a result, the new created vector $X_{new,i}^{HMS}$ has a better chance of achieving the global solution/knowledge.

This change can be mathematically expressed by Equation (10.34).

$$X_{new,i}^{HMS} = X_{old,i}^{HMS} + rand * \left(X_{teacher,i} - T_F X_{worst,i} \right) \qquad (10.34)$$

Where, $X_{worst,\,i}$ is the vector containing the population's worst fitness function.

10.5.3 Modifications in the teaching factor

T_F is randomly assigned a value of 1 or 2, implying that information transmitted from the perfect learner i.e., teacher to the leftover student is either 0% or 100%. However, transfer of information could range from 0% to 100% in practice. As a result, the T_F is changed from 0 to 1 to account for actual information exchange as shown below:

$$T_F = \left(\frac{1}{rand} \right)^a \qquad (10.35)$$

Where a is specified as the teaching factor rate. A big 'a' number assures a larger search space, increasing the chances of finding the global solution. Several case studies show that a value of 'a' between 0 and 5 ensures a global solution.

10.5.4 Modifications in the HS algorithm

The parameters HMCR, PAR, and BW are constant in basic HSA. To choose the values of HMCR and PAR, the suggested algorithm uses dynamic techniques [17]. In the hybrid algorithm, the dynamic selection of HMCR and PAR gives better balance in search space exploration and exploitation:

$$HMCR = HMCR_{max} - \left(HMCR_{max} - HMCR_{min} \right) * \left(\frac{current\ iter}{max\ iter} \right) \qquad (10.36)$$

$$PAR = PAR_{max} - \left(PAR_{max} - PAR_{min} \right) * \left(\frac{current\ iter}{max\ iter} \right) \qquad (10.37)$$

$$BW = BW_{max} * \exp \left[\ln \left(\frac{BW_{max}}{BW_{min}} \right) * \frac{current\ iter}{max\ iter} \right] \qquad (10.38)$$

Where $HMCR_{min}$ and $HMCR_{max}$ are the minima and maxima of HMCR values, respectively, PAR_{min} and PAR_{max} are the minima and maxima of PAR values, and BW_{min} and BW_{max} are the minimum and maximum BW values.

10.6 PROPOSED HS-TLBO ALGORITHM

The TLBO and HS algorithms have complementary properties; for example, the HS algorithm has outstanding exploratory performance but sluggish convergence, whereas TLBO has excellent exploitation features and converges rapidly. Hence the suggested method tries to combine the benefits of both techniques to swiftly arrive at a global optimum. So, for the most comprehensive exploration of the search process, HSA is used first, and then TLBO is used to utilize the search space. In the suggested algorithm, the selection of teaching phase in the TLBO algorithm and local pitch adjustment by HMCR in the harmony search algorithm is done in accordance with autoselection rates (ASR) which is expressed below:

$$ASR = \frac{Best\ fitness\ of\ HM}{Worst\ fitness\ of\ HM} \tag{10.39}$$

Because the magnitude of ASR update with iterations, it is complex, dynamic, and self-adaptive for selecting either of these two techniques for harmony vector improvisation. Because the best fitness value is always less than the worst fitness value in minimization situations, ASR will be less than 1. ASR is obtained for maximizing issues by reducing the inverse of the fitness values, resulting in ASR less than 1. Because ASR will initially be low, the HS method will be preferred for better search space exploration. As ASR approaches unity, the teaching phase is chosen to fine-tune the variables.

To mitigate the drawbacks of both HS and TLBO, the HTLBO first employs HS's exploring capabilities. After that, the TLBO algorithm is used to fine-tune the solution vector and speed up convergence. The HTLBO algorithm considers the following steps.

1. Target vector selection: $X_{new}^{HMS} = X_{old}^{HMS}$ is the harmony vector.
2. Generation of target vector $X_{new,i}^{HMS}$ in each iteration consisting of the following steps.

Step 1: Autoselection rate
        The ASR influence the selection of teaching phase or HMCR for the generation of the new target vector X_{new}^{HMS}. ASR is evaluated with the help of Equation (10.39).

Step 2: Teaching phase/HMCR phase
        The teaching phase is chosen when the magnitude of ASR is less than r, where value of r lies between 0 and 1 and $X_{new,i}^{HMS}$ is created using Equation (10.34). When ASR magnitude is more than r, the HS method is used to generate X_{new}^{HMS}. The HMCR determines the likelihood of tuning the target vector. If the random number is less than HMCR, Equation (10.37) is used to change the local pitch of

Table 10.1 Basic parameters of the HS-TLBO algorithm

Parameter	Values
a	0.2
HMCR$_{min}$	0.72
HMCR$_{max}$	0.95
PAR$_{min}$	0.35
PAR$_{max}$	0.55
r	0.9
BW$_{min}$	0.001
BW$_{max}$	0.1
Maxiter	100
HMS	90

$X_{new,i}^{HMS}$; otherwise, the mutation phase is used to construct the new target vector X_{new}^{HMS} in the feasible space using Equation (10.30). If the target vector X_{new}^{HMS} new has a higher fitness than the worst fitness vector, $X_{worst,i}$ is substituted by X^{HMS}.

Step 3: **Learning phase**

This phase interacts with the randomly picked harmony vector from harmony memory to improve the knowledge (fitness) of the target vector $X_{new,i}^{HMS}$. This is being improved in accordance with Equations (10.25) and (10.26). $X_{worst,i}$ is substituted by X_{new}^{HMS} if the fitness values of the target vector X_{new}^{HMS} is better than the fitness value of the worst fitness vector.

Step 4: **Dynamic adjustment of parameters**

The HMCR, PAR, and BW parameters are dynamically modified in each iteration for optimum functioning of the HTLBO algorithm, according to Equations (10.36)–(10.38).

The HM is updated if the target vector generated has a higher fitness; else, the next iteration begins. The algorithm is ended if the number of iterations meets the requirement for termination, and the best solution vector of HM is chosen as the objective function's solution.

Table 10.1 lists the parameters of the HTLBO algorithm. The flowchart in Figure 10.3 depicts the above steps of the proposed method in detail.

10.7 SIMULATION RESULTS AND MAIN FINDINGS

This section defines the execution of the anticipated algorithm, HS-TLBO, on 30 bus systems. The proposed HS-TLBO is used to locate and size EVPL

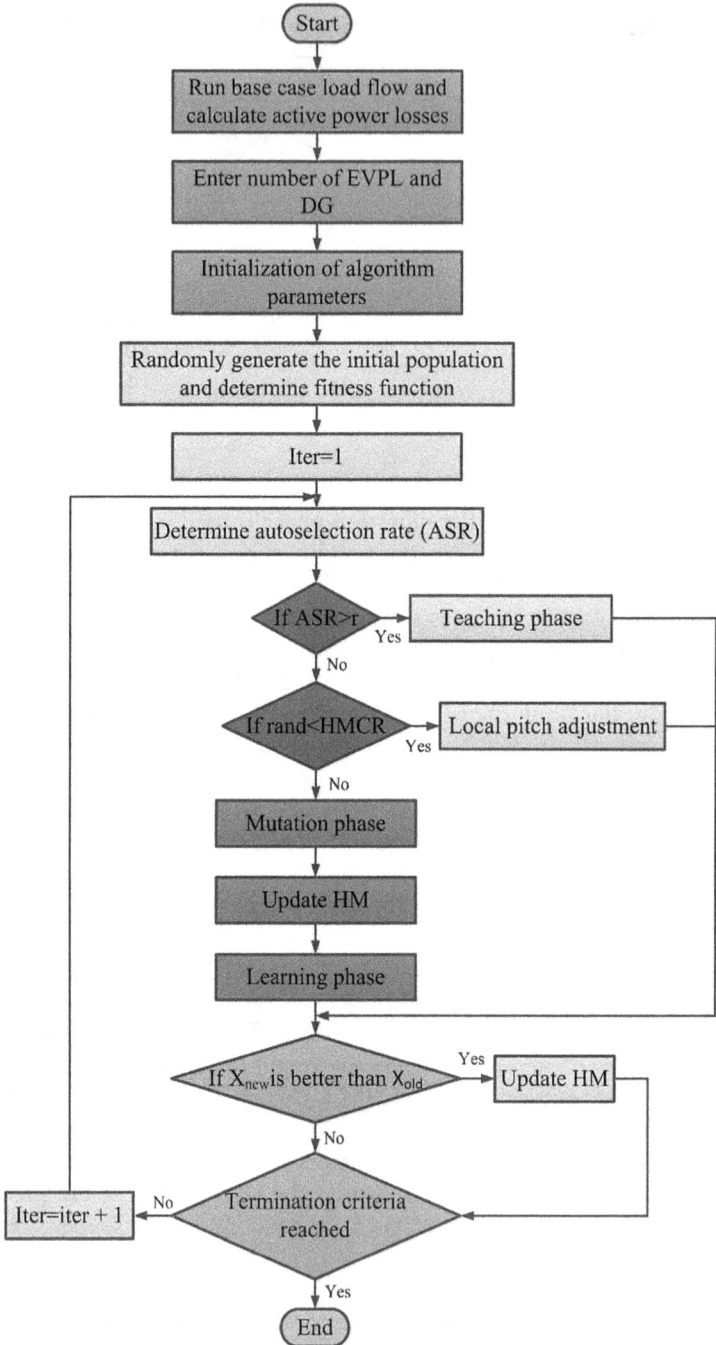

Figure 10.3 Flow chart of HS-TLBO algorithm.

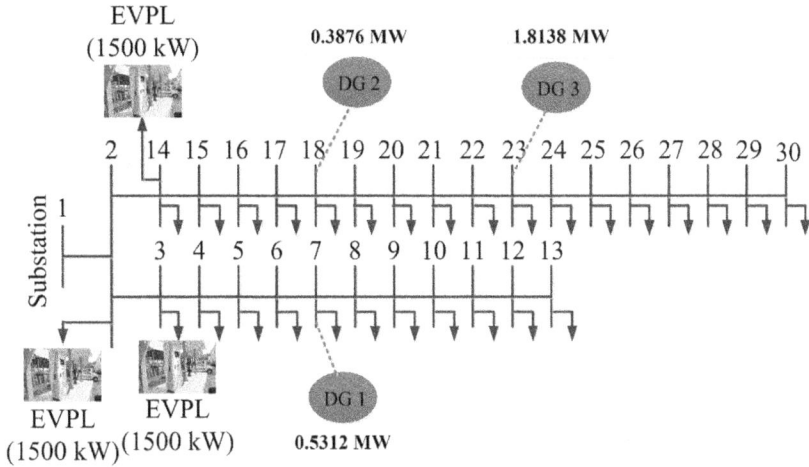

Figure 10.4 Configuration of 30-bus system with three EVPL and three DGs.

and DG units in 30-bus distribution systems. The outcomes are equated with TLBO for the objective function of: (1) reducing active power loss; and (2) reducing voltage deviation. The proposed HS-TLBO method is implemented in MATLAB R2015a on a desktop PC with a 3.2 GHz Intel i7 processor and 4 GB of RAM.

Figure 10.4 depicts a complete diagram of a 30-bus distribution network with three DG and three EVPL. The network of 30 bus is made up of 30 nodes and 29 lines. The system is capable of 100 kVA and 12.66 kV operation. It has a total reactive power load of 14248.26 kVAr and a total real power load of 22483.25 kW. Each EVPL is supposed to have 30 ports for charging, each with a capacity of 50 kW. Hence, it is possible to serve 30 EVs at once. In the distribution network, the optimum number of EVPL must be sited at the optimal node. Because EVPL installation increases network losses. Therefore, DGs are perfectly positioned to compensate for EVPL losses. The proposed HS-TLBO method is used to minimize power loss. A direct approach for load flow analysis is performed prior to the installation of EVPL and DG to find the initial losses of the system. The basic parameters of the anticipated technique, i.e., HS-TLBO, is tabulated in Table 10.1. Before installing EVPL and DGs, the active and reactive power values were found to be 607.6 kW and 273.6 kVAr, respectively. Furthermore, a minimum voltage of 0.9598 p.u. is indicated at bus 29.

Owing to enhanced EV loading, the inclusion of EVPL in the distribution network accelerates the active power losses while simultaneously lowering the voltage profile. As a consequence, the EVPL should be assigned as efficiently as possible, with as little increase in active power losses as possible. It's important to note that integrating a fixed - sized EVPL to bus 2 results in a 634.3 kW power loss. In order to address power loss issues and meet customer demand, a growing number of EVPL is essential to meet customer demand while also ensuring EVPL availability for EV users. The optimal placement of the second EVPL on bus 3 results in a 644.5 kW active power loss. In addition, when the third EVPL is optimally installed at bus 14, this results in a 652.5 kW active power loss.

This work addresses seven different scenarios in order to validate the methodology. The scenarios are as follows:

Scenario 1:   30-bus distribution network with no extra load i.e., already connected load only
Scenario 2:   Only one EVPL in a distribution network
Scenario 3:   Two EVPLs in a distribution network
Scenario 4:   Three EVPLs in a distribution network
Scenario 5:   The addition of one DG unit in a distribution network
Scenario 6:   The addition of two DGs
Scenario 7:   The addition of three DGs

When DG is positioned in an appropriate location and of appropriate size, active power losses is lowered and the voltage profile is enhanced. Because I^2R losses predominate, a lot of research articles have aimed at reducing active power losses in the electricity system. The active power losses and VDI are calculated prior to the reliability evaluation.

To lessen the charging effect of EVs, three DGs have been installed in the 30-bus network. The active power loss is 415.1 kW when one 0.5312 MW DG is optimally located at bus 7. The power loss drops to 307.5 kW when two DG units of sizes 0.3876 MW and 1.8138 MW are optimally situated at buses 18 and 23, respectively, in the grid network. Similarly, power loss further decreases to 237.9 kW when three DG units of sizes 0.5312 MW, 0.3876 MW, and 1.8138 MW are optimally placed at buses 7, 18, and 23,

Table 10.2 Active power loss values after incorporating EVPL and DG using HS-TLBO

Scenarios	Bus number and size	Active power loss (kW)
Base case	—	607.6
1 EVPL	1500 kW at bus 2	634.3
2 EVPL	1500 kW at bus 2 and 3	644.5
3 EVPL	1500 kW at bus 2, 3 and 14	652.5
1 DG	1.8138 MW at bus 23	415.1
2 DGs	0.3876 MW and 1.8138 MW at bus 18 and 23	307.5
3 DGs	0.5312 MW, 0.3876 MW and 1.8138 MW at bus 7, 18 and 23	237.9

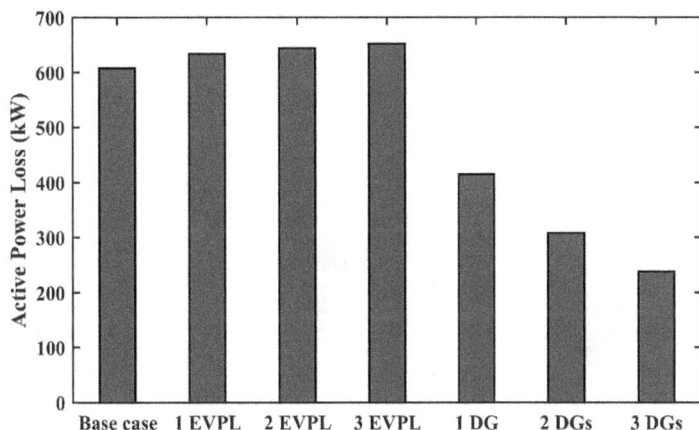

Figure 10.5 The effect of EVPL and DG incorporation on system's loss in a 30-bus system.

respectively. Table 10.2 displays the active power loss values when EVPL and DGs are installed sequentially in 30 bus distribution networks.

Figure 10.5 depicts the change in power loss after the setting up of EVPL and DG units. The required size of EVPL and DG, and also their implication on power loss, are compared in Table 10.2. When the same parameters are used, the results obtained by implementing HS-TLBO outperform TLBO, as shown in Table 10.3.

Table 10.3 Size, location and active power losses obtained by applying HS-TLBO and TLBO

	HS-TLBO				TLBO			
		Optimal DG		Active power loss (kW)		Optimal DG		Active power loss (kW)
Scenarios	EVPL location	location	Size (MW)		EVPL location	location	Size (MW)	
Base case	–	–	–	607.6	–	–	–	609.8
1 EVPL	2	–	–	634.3	2	–	–	638.5
2 EVPL	2, 3	–	–	644.5	2, 3	–	–	647.3
3 EVPL	2, 3, 14	–	–	652.5	2, 3, 14	–	–	655.1
1 DG	2, 3, 14	23	1.8138	415.1	2, 3, 14	23	1.8245	420.5
2 DGs	2, 3, 14	18	0.3876	307.5	2, 3, 14	18	0.4087	310.1
		23	1.8138			23	1.8245	
3 DGs	2, 3, 14	7	0.5312	237.9	2, 3, 14	7	0.5435	241.4
		18	0.3876			18	0.3845	
		23	1.8138			23	1.8154	

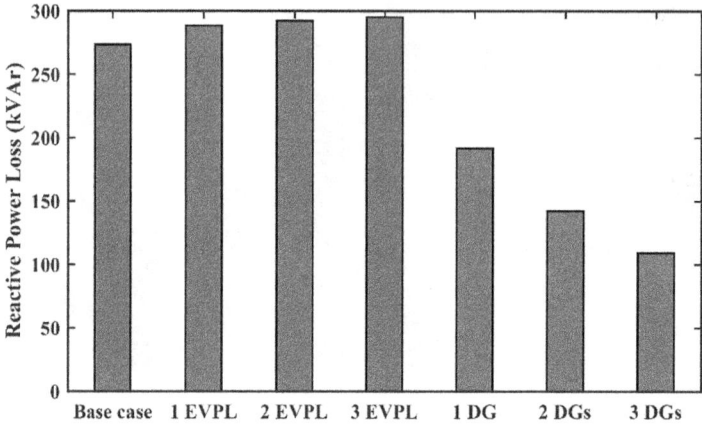

Figure 10.6 The effect of EVPL and DG integration on reactive power loss in a 30-bus system.

DG placement, like active power loss, reduces reactive power loss, as shown in Figure 10.6. The reactive power loss in the base case was 273.6 kVAr. It rises when EVPL are deployed in the network. However, with the addition of three DGs, it drops to 109.2 kVAr. Figure 10.6 depicts the varying behaviour of reactive power losses.

Similarly, the proposed technique, HS-TLBO, is validated by equating the gained outcomes to those of other prevailing methods, such as TLBO.

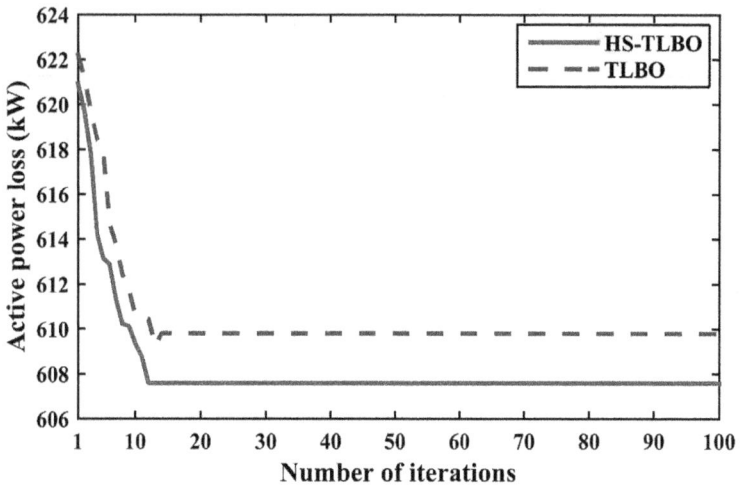

Figure 10.7 Response of HS-TLBO and TLBO for active power loss in 30-bus network.

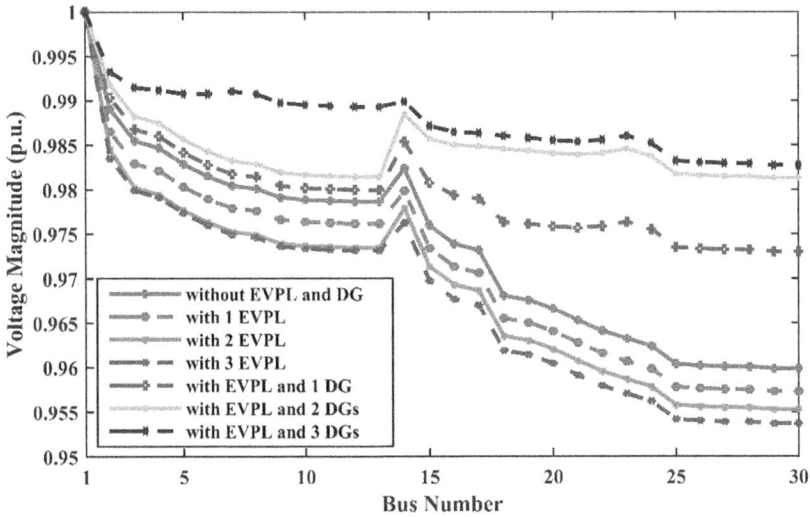

Figure 10.8 Effect of EVPL and DG integration on voltage profile of 30-bus system.

The HS-TLBO method has a lower active power loss, i.e., 237.9 kW than TLBO (239.5 kW). The convergence of active power loss over iteration using the proposed HS-TLBO and TLBO is depicted in Figure 10.7. Furthermore, the convergence properties show that HS-TLBO reaches optimum solution quicker than isolated TLBO.

The inclusion of EVPL has an adverse influence on the voltage profile, as with system loss. As EV loading increases, the system's voltage profile deteriorates. These disturbances are reduced by incorporating DG units at proper distribution network nodes. When several EVPL and DGs are placed, the voltage profile of the 30-bus system is depicted in Figure 10.8.

DGs and EVPL are incorporated to ensure the smooth running of the distribution system. The incorporation of DG units improves the system's voltage levels. The improved voltage profile after the addition of DG units is depicted in Figure 10.8. Voltages on all buses fluctuate due to actual and reactive power losses in the distribution system. Hence, actual power assistance is needed to minimize power losses while increasing voltage levels by decreasing I^2R losses. When a large number of DGs are located, bus voltages have also been observed to improve.

The minimum magnitude of voltage for one DG is 0.9729 p.u at bus number 30; the minimum magnitude of voltage in case of two DG units is 0.9812 p.u at bus number 30 by contrast, the minimum voltage appears at bus 30 of magnitude 0.9827 p.u. when three DGs are optimally placed. Hence, it is possible to conclude that using multiple DGs improves minimum voltage.

Figure 10.9 depicts the flow of active power in all lines of the 30-bus distribution system after EVPL and DGs are added. As EVPL loading increases,

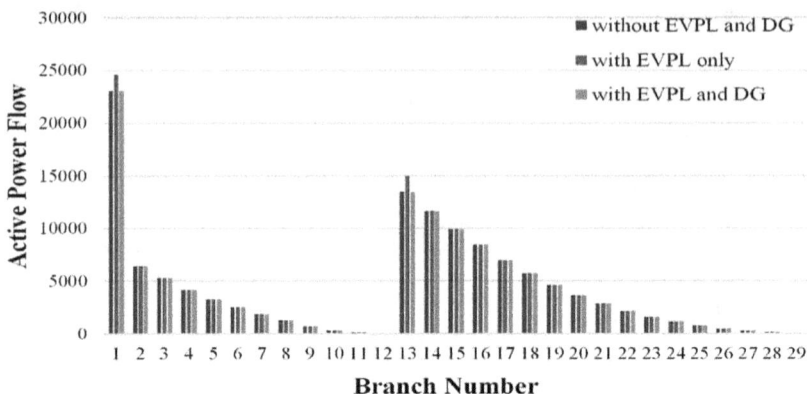

Figure 10.9 Impact of EVPL and DG placement on active power flow in 30-bus system.

active power flow increases, but it is regulated by placing DGs at appropriate nodes and keeping it within permitted limits.

10.8 CONCLUSION

EVs are a promising option for reducing pollution caused by transportation. EVPL have grown in popularity as EVs become increasingly widespread; nonetheless, the negative impact of the EV charging station loads on the electrical grid should not be underestimated. Using a direct approach-based load flow analysis, this chapter investigates the impact of EVPL on the IEEE standard system. More grid power is required to charge electric vehicles, resulting in increased power losses. As a result, to compensate for EVPL power losses, DG should be used. To compensate for the system's power loss, this study uses Type 2 DG. HS-TLBO, a hybrid algorithm, was also used to reduce losses by determining the best node for EVPL and DG placement.

On 30-bus systems, the proposed hybrid algorithm has been tested. The accuracy of the established method is further evaluated by comparing results to other methodological procedures, such as TLBO. For a 30-bus system, HS-TLBO shows a significant decrease in system losses when compared to TLBO.

The number of EV charging loads is fixed, and DGs are added to the grid network to reduce system losses, aside from voltage and current limits. Power engineers may easily determine the number of DGs to compensate for the effect of EVPL by studying the capacity mismatch between the increased EVPL demand and overall system load. Three DGs considerably improve the performance of the 30-bus system. Increasing the number of DGs reduces power losses and improves the voltage profile. However, a fourth DG has minimal effect.

REFERENCES

[1] Z. Liu, F. Wen, and G. Ledwich, "Optimal planning of electric-vehicle charging stations in distribution systems," *IEEE Trans. Power Deliv.*, vol. 28, no. 1, pp. 102–110, Jan. 2013, doi:10.1109/TPWRD.2012.2223489

[2] H. R. Galiveeti, A. K. Goswami, and N. B. Dev Choudhury, "Impact of plug-in electric vehicles and distributed generation on reliability of distribution systems," *Eng. Sci. Technol. an Int. J.*, vol. 21, no. 1, pp. 50–59, Feb. 2018, doi:10.1016/j.jestch.2018.01.005

[3] H. Mehrjerdi and E. Rakhshani, "Vehicle-to-grid technology for cost reduction and uncertainty management integrated with solar power," *J. Clean. Prod.*, vol. 229, pp. 463–469, Aug. 2019, doi:10.1016/j.jclepro.2019.05.023

[4] M. Tomasov, D. Motyka, M. Kajanova, and P. Bracinik, "Modelling effects of the distributed generation supporting e-mobility on the operation of the distribution power network," *Transp. Res. Procedia*, vol. 40, pp. 556–563, 2019, doi:10.1016/j.trpro.2019.07.080

[5] W. Kempton and J. Tomić, "Vehicle-to-grid power implementation: From stabilizing the grid to supporting large-scale renewable energy," *J. Power Sources*, vol. 144, no. 1, pp. 280–294, Jun. 2005, doi:10.1016/j.jpowsour.2004.12.022

[6] D. C. Crowley, "Comparative environmental analysis of both plug-in hybrid and battery electric cars in Azerbaijan," 2020, doi:10.13140/RG.2.2.34695.55207

[7] S. Faddel, A. Al-Awami, and O. Mohammed, "Charge control and operation of electric vehicles in power grids: a review," *Energies*, vol. 11, no. 4, p. 701, Mar. 2018, doi:10.3390/en11040701

[8] Z. Darabi and M. Ferdowsi, "Aggregated impact of plug-in hybrid electric vehicles on electricity demand profile," *IEEE Trans. Sustain. Energy*, vol. 2, no. 4, pp. 501–508, Oct. 2011, doi:10.1109/TSTE.2011.2158123

[9] S. Shafiee, M. Fotuhi-Firuzabad, and M. Rastegar, "Investigating the impacts of plug-in hybrid electric vehicles on power distribution systems," *IEEE Trans. Smart Grid*, vol. 4, no. 3, pp. 1351–1360, Sep. 2013, doi:10.1109/TSG.2013.2251483

[10] M. K. Gray and W. G. Morsi, "On the impact of single-phase plug-in electric vehicles charging and rooftop solar photovoltaic on distribution transformer aging," *Electr. Power Syst. Res.*, vol. 148, pp. 202–209, Jul. 2017, doi:10.1016/j.epsr.2017.03.022

[11] K. Nekooei, M. M. Farsangi, H. Nezamabadi-Pour, and K. Y. Lee, "An improved multi-objective harmony search for optimal placement of DGs in distribution systems," *IEEE Trans. Smart Grid*, vol. 4, no. 1, pp. 557–567, Mar. 2013, doi:10.1109/TSG.2012.2237420

[12] Jen-Hao Teng, "A direct approach for distribution system load flow solutions," *IEEE Trans. Power Deliv.*, vol. 18, no. 3, pp. 882–887, Jul. 2003, doi:10.1109/TPWRD.2003.813818.

[13] M. S. K. Reddy and K. Selvajyothi, "Optimal placement of electric vehicle charging station for unbalanced radial distribution systems," *Energy Sources, Part A Recover. Util. Environ. Eff.*, pp. 1–15, Feb. 2020, doi:10.1080/15567036.2020.1731017

[14] D. P. Kothari, "Power system optimization," in *2012 2nd National Conference on Computational Intelligence and Signal Processing (CISP)*, Mar. 2012, pp. 18–21, doi:10.1109/NCCISP.2012.6189669

[15] S. Sultana and P. K. Roy, "Multi-objective quasi-oppositional teaching learning based optimization for optimal location of distributed generator in radial distribution systems," *Int. J. Electr. Power Energy Syst.*, vol. 63, pp. 534–545, Dec. 2014, doi:10.1016/j.ijepes.2014.06.031

[16] R. V. Rao, V. J. Savsani, and D. P. Vakharia, "Teaching–learning-based optimization: A novel method for constrained mechanical design optimization problems," *Comput. Des.*, vol. 43, no. 3, pp. 303–315, Mar. 2011, doi:10.1016/j.cad.2010.12.015

[17] P. Yadav, R. Kumar, S. K. Panda, and C. S. Chang, "An Intelligent Tuned Harmony Search algorithm for optimisation," *Inf. Sci. (Ny).*, vol. 196, pp. 47–72, Aug. 2012, doi:10.1016/j.ins.2011.12.035

Chapter 11

An intelligent technique for electric vehicles for the monitoring of parameters

Anjali Jain and Ashish Mani
Amity University, Noida, India

Anwar S. Siddiqui
Jamia Milia Islamia, New Delhi, India

CONTENTS

11.1 INTRODUCTION

Plug-in electric vehicles offer numerous advantages to the planet as well as to the general population who drive them. Among those advantages are less pollution, and running costs which are much lower than fossil fuel-consuming vehicles. Plug-in electric vehicles have become very popular nowadays, yet there are several issues associated with electric vehicles for the utility are: uncertainty in various parameters such as the driving pattern of users; the travelling time of plug-in electric vehicles; and idling time of the plug-in vehicle in which it can be connected to the grid as a source. Other issues include: a need for information about the electric vehicles' charging time;, the continuous monitoring of electric vehicles by the user; charge scheduling; the tariff for charging the electric vehicle, and so on.

The principal object of the work is to propose an idea that will provide an electric vehicle with a Global Positioning System (GPS) and a Battery Management System (BMS) to trace the location of a plug-in electric vehicle with time stamping and monitor other parameters for the optimization of

DOI: 10.1201/9781003311195-11

location, scheduling, forecasting and tariff of power being sold and purchased, etc. In addition, the proposed work will help the utility to construct the profiling of the consumption of electrical energy, run a scheduling algorithm for the charging needs of users with more effectiveness, the installation of new charging stations, the ability to use the surplus power of the electric vehicle in the vehicle to grid (V2G) mode, and advice to EV users for their vehicles to be charged under the most economical tariff.

This section introduces the outline of the project and the objectives to be achieved. Section 11.2 summarizes the associated literature survey. Section 11.3 proposes the conceptualization of the work. Section 11.4 talks about the flow of information and hardware blocks are discussed in Section 11.5.

11.2 LITERATURE SURVEY

This section discusses the different state-of-the-art work which has been done in this field to date. In [1] there is a discussion of the use of a BMS to monitor the key parameters of batteries, such as current, voltage and temperature. It also monitors the health of cells in the battery; if there is any issue, an alarm is initiated. Chacko and Sachidanandan developed an intelligent energy management system for PHEV plug-in hybrid electric vehicles grid integration. The purpose of their work is to optimize load sharing between the power trains of PHEV [2]. In [3] a predictive intelligent battery management system is proposed which contributes to the optimization of energy efficiency and reduced emissions. In [4], a GPS/GSM module is used to monitor and manage the EV battery. In addition, an Android application is used in this work.

Reference [5] provides a detailed report on different Li-ion batteries and also explains the fundamentals, structures and overall performance evaluation of different batteries. It also demonstrates different features, such as cell condition monitoring, charge, and discharge control, etc. and also aims to enhance the overall performance of the system. In [6], thermal management, battery equalization and fault diagnostic are studied and investigated. Then, in [7] the need for a BMS is discussed; this work also highlighted that a BMS should contain an accurate algorithm to estimate the condition of the battery.

Reference [8] relates to systems and methods may enable customers to reserve time slots at charging stations for recharging electric vehicles. The systems and methods may include receiving, from a customer's computing device, a reservation inquiry to access a charging station to recharge an EV providing, to the customer computing device, a respective cost for accessing the charging station during one or more available time periods; receiving a selection from the customer's computing device for at least a portion of the available time periods; and delivering, to the customer computing device, a reservation confirmation for the selected time period. whereas the proposed work gives information with regard to a travelling pattern, travelling time, and idling time so that the utility can maintain a better load profile.

Reference [9] relates to a travelable distance calculation system for a vehicle includes a travel history database that stores primary data (D1), which includes the electricity consumption history of a target vehicle, and secondary data (D2), which includes an electricity consumption history of a plurality of vehicles other than the target vehicle; and an arithmetic unit configured to calculate the travelable distance of the target vehicle using at least one of the primary data (D1) and the secondary data (D2). As per [9], travel distance is calculated for EV whereas in the proposed work not only travel distance is calculated but also travel profile is also obtained and the statistics of the travelling pattern along with travel time and idle time is utilized by utility for better and smoother load profile.

11.3 CONCEPTUALIZATION OF WORK

The proposed work relates to an electric vehicle with a Global Positioning System (GPS) and a Battery Management System (BMS) wherein the introduced GPS traces the location of a plug-in electric vehicle with time stamping. Traveling period, idling period, charging period, the status of charge, and preferred charging time of electric vehicle are sent to the utility using BMS. These help the utility to construct the profiling for the consumption of electrical energy and help to run a scheduling algorithm for the charging needs of users with more effectiveness.

Figure 11.1 shows the GPS and BMS module attached to 'n' electric vehicles. GPS will communicate with the satellite wherein the information about the electric vehicle, consisting of the route it is taken, will be sent with the help of GPS. The other information, such as traveling time, idling time, plug-in or charging status, battery status, and preferred time of charging, will be sent by BMS to the Central Processing Unit (CPU) of each electric vehicle and will be operated using a Human Machine Interface (HMI) wherein the GPS module will be sending the real-time location along with time stamping. This information is processed in the CPU and the required information will be sent to the network using the suitable network interface. This information from the network will be sent to the data centre server and then to the network cloud which can also share different power system information about load and spinning reserves from the utility, the need for charging stations and their distance from charging station owners. They also address the need for electric vehicles with market players such as EV manufacturing units, and also to EV users for giving guidance for the charging of EVs with the most economical tariffs.

The information received by the utility, charging station owners, and other market players will help to find what is the traveling time of individual

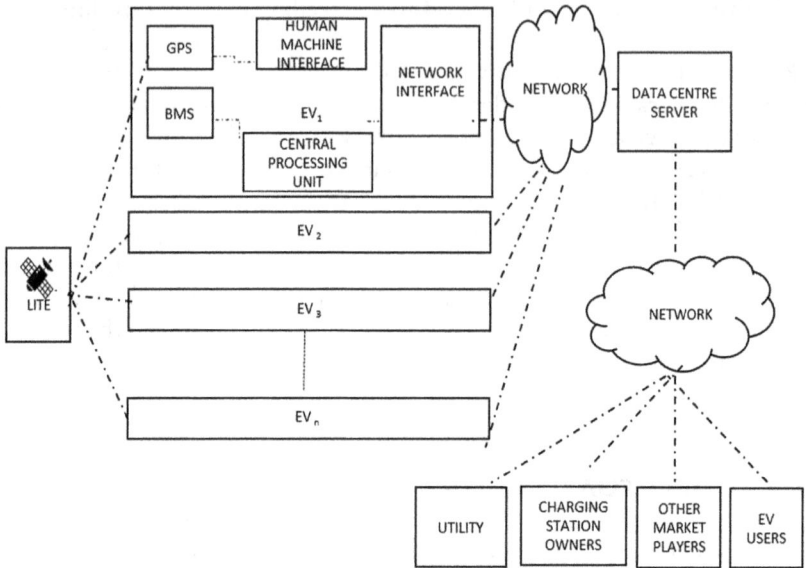

Figure 11.1 Block diagram for monitoring of parameters.

electric vehicles, along with their idling time. It also encourages EV owners to work in the V2G mode for which they will also get incentives and plan the commissioning of a charging station at the most suitable location. This will reduce the commissioning of spinning reserves, run the charging algorithm more efficiently with real-time data, and help in making accurate predictions with regard to power demand. Thus, it helps in deciding the tariff of power being sold and purchased.

Thus, the user receives a notification about preferred time for charging as per utility along with incentives, a nearby charging station for charging the vehicle, voltage available at the charger, and proposed charging time.

The proposed work will be very helpful to the user as well as the utility, charging station owners, and other market players in meeting the benefits of their interest.

Figure 11.2 represents the two-way communication of electric vehicles, charging stations, utility/grid, other market players' management systems, etc. with the system. Different parameters of the EV such as the traveling period, the distance from charging stations, the location of a vehicle, the state of charge, and the preferred charging time is sent to the central system with the help of a communication system wherein the central system is sending suitable information about the nearby charging station, and the tariff of charging and discharging at that timestamp to the EV, allowing informed decisions to be taken by the user.

Similarly, there also exists two-way communication between the central system and the charging station which helps the charging station to send the

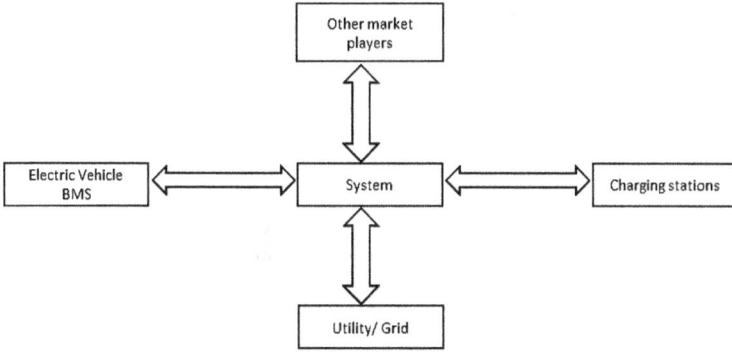

Figure 11.2 Block diagram for connectivity of different blocks.

information about waiting time, the availability of a charging station at the requested time stamp, and to also receive the information about the approaching EVs. The Central System, in turn, is also maintaining the profile of EVs, which will help the utilities/grid in planning new charging stations. Similarly, the information is also conveyed to other market players which are associated with EVs such as manufacturers, sellers, etc. The utility/grid sends information about the need for spinning reserves, tariffs to the central system; on the other hand, information about traffic congestion, the planning of new charging stations, etc. is communicated by other market players.

11.4 FLOW OF INFORMATION

The information flowchart is shown in Figure 11.3 and is explained in the section which follows.

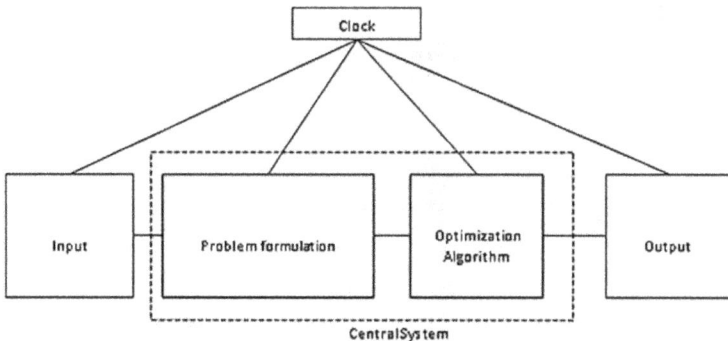

Figure 11.3 Block diagram for the flow of different information.

11.4.1 Input to central system

The input from the four different stakeholders, i.e., electric vehicles, charging stations, utility/grid, other market players, as shown in Figure 11.2, is taken and is communicated to the system with the help of a suitable communication medium. The data recorded need to be synchronized and hence a clock is utilized in order to fetch the information at the same instant. The information received from different input sources is formulated in the form of a multi-objective dynamic optimum problem, which is primarily used to solve scheduling and resource allocation while taking into account the constraints of different objectives. After processing, the output will be communicated to different stakeholders for the necessary action to be taken.

Figure 11.4 represents the input received from different stakeholders and maintained in the database of the central system. The real-time data is fetched from different stakeholders at the same timestamp with the help of the synchronizing clock. The inputs which are received from EVs are the depth of discharge, the state of charge, the battery type, and the location of the charging station tracked by GPS. The inputs communicated by charging stations to the central system are the availability of power, waiting for the time for EVs, the type of charging (fast or slow), the cost of charging, and location of the charging station. The grid operators also tell about the current tariff charges, the need of spinning reserves, the profile

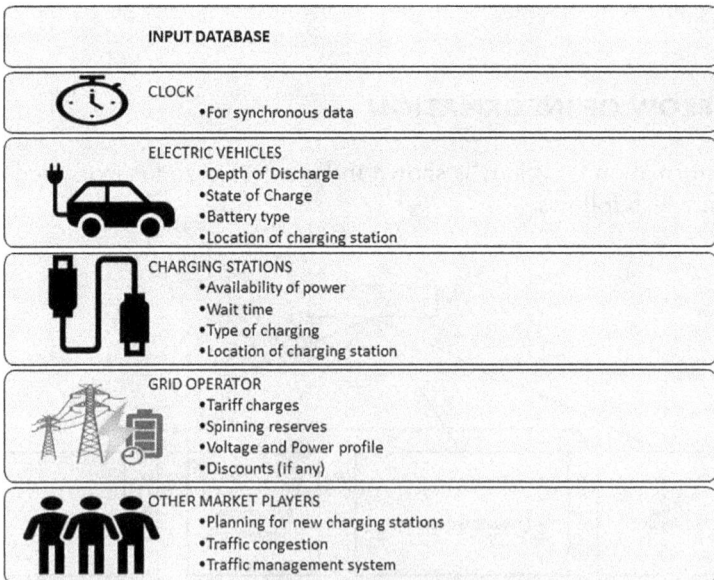

INPUT DATABASE

CLOCK
  • For synchronous data

ELECTRIC VEHICLES
  • Depth of Discharge
  • State of Charge
  • Battery type
  • Location of charging station

CHARGING STATIONS
  • Availability of power
  • Wait time
  • Type of charging
  • Location of charging station

GRID OPERATOR
  • Tariff charges
  • Spinning reserves
  • Voltage and Power profile
  • Discounts (if any)

OTHER MARKET PLAYERS
  • Planning for new charging stations
  • Traffic congestion
  • Traffic management system

Figure 11.4 Database for input block from different stakeholders.

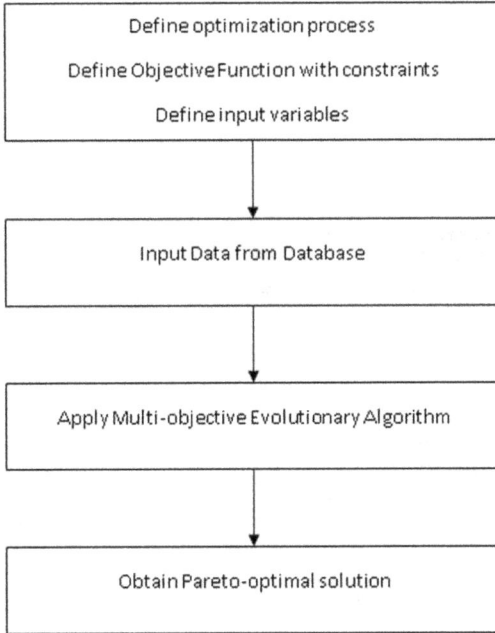

Figure 11.5 Flowchart for algorithm.

of voltage and power, and any discounts offered in the case of surplus power. Other market players will be sending inputs about traffic congestion on different routes, traffic management systems, the planning of new charging stations, etc.

11.4.2 Optimization process and algorithm

The different inputs received by different stakeholders are formulated in the form of a multi-objective dynamic optimization process whose main objective is to optimally schedule EVs and the allocation of charging stations while taking care of the constraints in the system. The problem formulation and algorithm are processed in the central system. The pseudo-code and flowchart for the algorithm are as under:

Pseudocode:

```
Define optimization process.
Define objective function with constraints
Define input variables
Input data from database
Apply multi-objective evolutionary algorithm
Obtain Pareto-optimal solution
```

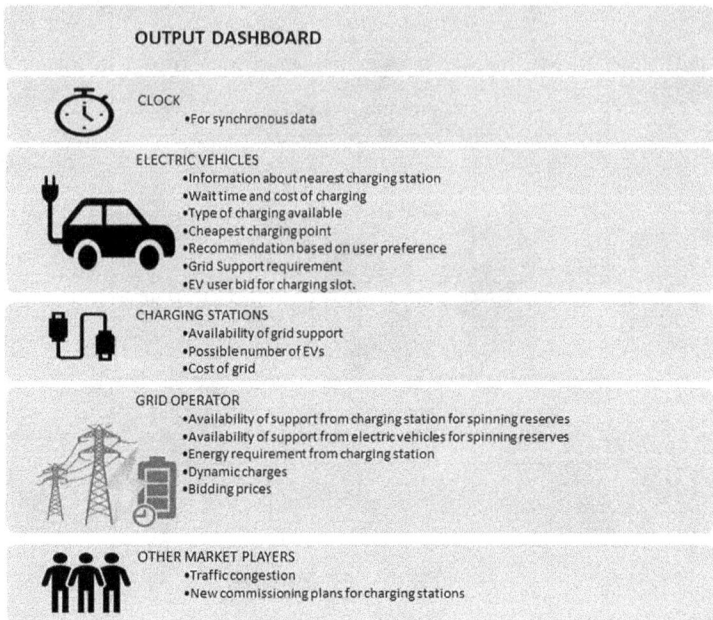

OUTPUT DASHBOARD

CLOCK
•For synchronous data

ELECTRIC VEHICLES
•Information about nearest charging station
•Wait time and cost of charging
•Type of charging available
•Cheapest charging point
•Recommendation based on user preference
•Grid Support requirement
•EV user bid for charging slot.

CHARGING STATIONS
•Availability of grid support
•Possible number of EVs
•Cost of grid

GRID OPERATOR
•Availability of support from charging station for spinning reserves
•Availability of support from electric vehicles for spinning reserves
•Energy requirement from charging station
•Dynamic charges
•Bidding prices

OTHER MARKET PLAYERS
•Traffic congestion
•New commissioning plans for charging stations

Figure 11.6 The output dashboard for sending information to different stakeholders.

11.4.3 Dashboard for output

Figure 11.6 represents the output that is shown on the dashboard. The information which is communicated by central servers to all the stakeholders is showcased. The clock is set to send synchronous data. The output received by electric vehicles from the central system is information about the nearest charging station, the waiting time at the station, the cost of charging for the required charging, the type of charging available (whether fast or slow), the cheapest charging station (which may be at some distance), and the recommendation by the central system to the EV used based on the historical data available in the database. The central system also shares the requirement of the grid from EV users in instances in which the user has a surplus of power, meaning that they can bid for charging slots.

The charging station receives information about the availability of grid support, the number of EVs approaching the station, and tariffs of the grid for charging. The grid operators may benefit from getting information about the support available from charging stations and EVs, energy requirements from charging stations, and dynamic charges in case someone needs to charge urgently.

Other market players receive information about traffic congestion after the running of algorithms and what should be done regarding commissioning plans for new stations.

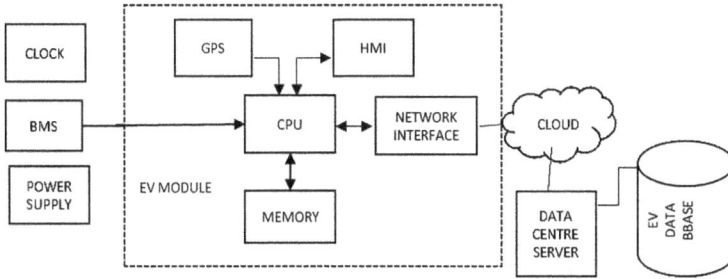

Figure 11.7 Hardware blocks for electric vehicles.

Following the discussion of information blocks in this section, the following section discusses the hardware blocks for different stakeholders.

11.5 HARDWARE BLOCKS

Figure 11.7 represents the hardware block for an EV wherein the inputs for the Central Processing Unit are from a BMS, a GPS, the Human–Machine Interface, and Memory. The synchronizing clock is used for getting the time stamp of the information and the auxiliary power supply is also used for the continuous flow of information. The CPU is interfaced to the cloud, and the information from different such electric vehicles is sent to the Data Centre Server and the database is maintained as shown in Figure 11.4.

Figure 11.8 represents the hardware block for charging stations wherein the charging station is connected to the Central Processing Unit with an interface. The CPU is also connected to the HMI and Memory. The clock is also used to get the time stamp of the information and the auxiliary power supply is also used for the continuous flow of information. The CPU is interfaced to the cloud where the information about different charging stations is sent to the Data Centre Server and the database is maintained in the charging stations database, as shown in Figure 11.4.

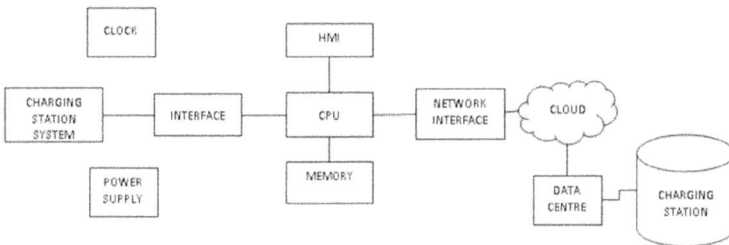

Figure 11.8 Hardware blocks for charging stations.

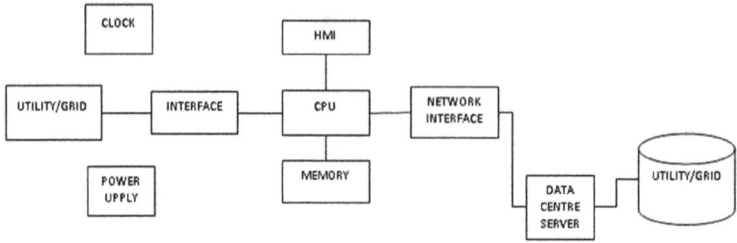

Figure 11.9 Hardware blocks for Utility/Grid.

Figure 11.9 represents the hardware block for the utility/grid wherein the utility/grid is connected to the Central Processing Unit with an interface. The CPU is also connected to HMI and the Memory. The clock is used for getting the time stamp of the information and the auxiliary power supply is also used for the continuous flow of information. The CPU is interfaced to the Data Centre server and the database is maintained in charging stations database as shown in Figure 11.4.

Figure 11.10 represents the hardware block for different market players wherein the market player is connected to the Central Processing Unit with an interface. The CPU is also connected to the HMI and the Memory. The clock is used for getting the time stamp of the information and the auxiliary power supply is also used for the continuous flow of information. The CPU is interfaced to the cloud where the information about such different market players is sent to the Data Centre Server and the database is maintained in the market players' database, as shown in Figure 11.4.

Figure 11.11 shows that the information is collected from the GPS, the BMS, and the user and is sent to local memory with time stamping. This process is repeated at regular time intervals. The Data Centre Server will store information in different stakeholders' databases. After processing the information, the information is sent back to the stakeholders, as is displayed in the HMI/Dashboard at different stakeholders' places.

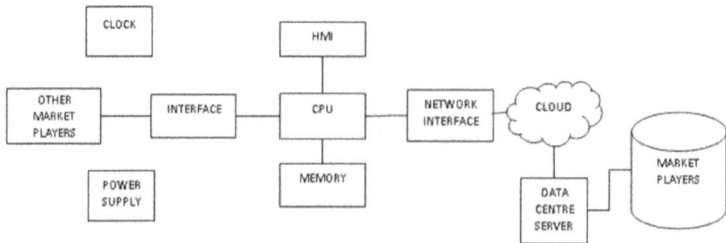

Figure 11.10 Hardware blocks for Other Market Players.

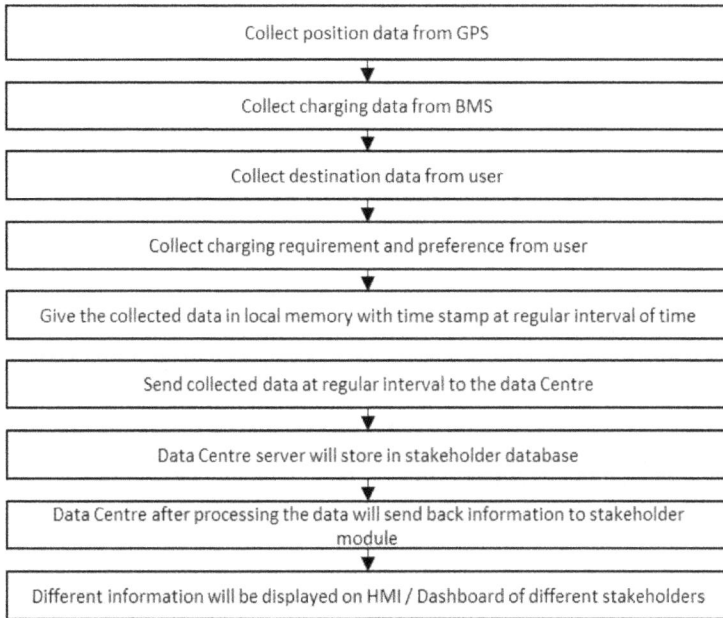

Figure 11.11 Detailed flowchart for the proposed system.

11.6 CONCLUSION

A GPS- and BMS-enabled electric vehicle proposed in this work can be utilized to monitor parameters to provide an EV with GPS to trace the location of a plug-in electric vehicle with time stamping. It also provides a BMS to monitor other parameters for the optimization of location, and the scheduling of EVs. It can also be used to provide forecasting and the tariff for power being sold and purchased, etc. One of the purposes of this work is to help the utility to construct the profiling of the consumption of electrical energy. It also helps to run scheduling algorithms for the charging needs of users with greater effectiveness. The proposed scheme also helps the market stakeholders to find the profiling of EVs which will help in the siting of new charging stations. Finally, the system proposed will also guide EV users for charging EVs with economical tariffs.

The future scope of work will include the implementation of the proposed scheme in a real-time system.

REFERENCES

[1] Sivaraman, P., and C. Sharmeela. "IoT-based battery management system for hybrid electric vehicle." *Artificial Intelligent Techniques for Electric and Hybrid Electric Vehicles* (2020): 1–16.

[2] Chacko, Parag Jose, and Meikandasivam Sachidanandam. "Optimization & validation of intelligent energy management system for pseudo dynamic predictive regulation of plug-in hybrid electric vehicle as donor clients." *eTransportation* 3 (2020): 1–11.

[3] Abdul-Hak, Mohamad, Nizar Al-Holou, and Utayba Mohammad. "Predictive intelligent battery management system to enhance the performance of electric vehicle." *Electric Vehicles–Modelling and Simulations* (2011): 365–384.

[4] Gupta, Karan, and Vilas H. Gaidhane. "Smart IOT-enabled battery management system for electric vehicle." In *Proceedings of International Conference on Communication and Artificial Intelligence*, pp. 153–163. Springer, Singapore, 2021.

[5] Hannan, Mahammad A., Md Murshadul Hoque, Aini Hussain, Yushaizad Yusof, and Pin Jern Ker. "State-of-the-art and energy management system of lithium-ion batteries in electric vehicle applications: Issues and recommendations." *IEEE Access* 6 (2018): 19362–19378.

[6] Lipu, M. S. Hossain, M. A. Hannan, Tahia F. Karim, Aini Hussain, Mohamad H. M. Saad, Afida Ayob, Md Sazal Miah, and T.M. Indra Mahlia. "Intelligent algorithms and control strategies for battery management system in electric vehicles: Progress, challenges and future outlook." *Journal of Cleaner Production* 292 (2021): 126044. 1–27.

[7] Rahimi-Eichi, Habiballah, Unnati Ojha, Federico Baronti, and Mo-Yuen Chow. "Battery management system: An overview of its application in the smart grid and electric vehicles." *IEEE Industrial Electronics Magazine* 7, no. 2 (2013): 4–16.

[8] Patent: Application No. 1513/DEL/2012 A INDIA Date of filing of Application: 17/05/2012 Publication Date : 04/12/2015 Title of the invention: System And Methods For Reservations Of Charging Stations For Electric Vehicles.

[9] Patent: Application No. 201714025828 A INDIA Date of filing of Application: 20/07/2017 Publication Date: 02/02/2018 Title of the invention: Travelable Distance Calculation System And Travelable Distance Calculation Method For Vehicle.

Chapter 12

Operational and economical aspects of a smart grid with large penetration of RESs and EVs

Anita Devi Ningthoujam
Manipur Technical University, Imphal, India

Divya Asija and Rajkumar Viral
Amity University, Noida, India

CONTENTS

DOI: 10.1201/9781003311195-12

12.1 INTRODUCTION

Electric vehicles (EVs) are becoming an increasingly important market due to the increased environmental concerns. To compensate for both the limited range of energy and to achieve an effective grid inclusion, the energy domain is currently integrating this new market into their model. A smart grid is defined as an electricity network that integrates a suite of instruction, communication and other innovations which observes and controls the transport of electricity from all generation sources in order to meet the changing electricity demands of various end-users. Electrification plays a significant role in the decarbonization of road transportation. This would require major investment, in terms of advancing the development of vehicles, the electricity network and associated charging electric systems. As the use of electric vehicles (EVs) continues to increase, this new type of electricity load needs special care and management in order to reduce impact on peak electricity demand and, therefore, the cost of supply, given the feasibility that a huge proportion of motorists would aim to recharge their vehicle batteries during the evening [1–4]. The evening is typically a period of high demand as people usually consume electricity during evening hours upon their return from their work and school, using home appliances such as lighting and heating/cooling while much office and industrial equipment also continues to operate. However, recent technological progress in electricity distribution and load management, as the so-called "smart grids", promise to accelerate the integration of EVs into electricity load and to lower costs. Electric services have already begun to utilize smart-grid technologies in order to improve and manage commercial and household load using intelligent metering and communication systems to save energy, cut emissions and reduce peak loads. More extensive implementation would enable EV charging to be planned intelligently. In addition, it could – perhaps in terms of principle – empower the storage capacity of the batteries in EVs to be used as an additional source of power at times of maximum load; the excess charge in those batteries could be fed back into the network during the evening peak time and the battery restored during nighttime [5]. There may also be an extent to which the utilization of this storage potential to make up for the concern over electricity supply from varying renewable energy sources such as wind and solar. Thus, both smart girds and EVs could be mutually beneficial and thus makes forward investment possible in smart grids.

12.2 COMBINED OPERATION: BEST PRACTICES AND MAJOR ISSUES

Given the increased use of EVs across the world, the electricity system also faces several challenges like considerable growth in demand, and the shifting of load patterns, where the integration of some sources of supply and variability of renewables-based supply come into the picture. This is made possible with the help of smart-grid technologies in meeting these challenges by offering a cost-effective option. By doing so, it helps in achieving energy more efficiently in a more secure and sustainable way [5, 6].

They can enable and incentivize customers for adjusting their demand in real time in case of market or network changes, accommodate generating sources and storage options, optimize utilization as well as operating efficiency of generation, transmission and distribution assets and provides flexibility to unplanned outages and interferences of supply [4].

Among the important major drivers of the smart grid are renewable sources in the form of wind and solar power.

12.2.1 States of Electric Vehicles

Plug-in vehicles have different features which satisfy the drivers' needs in a manner similar to that of a conventional type, where charging can be carried out efficiently simply by plugging the EV into an external power supply. It also has large battery packs for charging from the power grid, which have higher efficiency and lower costs when compared to conventional Hybrid Electric Vehicles (HEVs). This feature results in greater power output with greater durability of the drive system, which makes it distinct from other conventional HEVs.

Electric vehicles can be classified into two types: hybrid plug-in EV (PHEV) and All EV (AEV). In the case of the AEV, the motors are driven electrically using some external sources. These can be further divided into battery EVs (BEVs) and fuel cell EVs (FCEVs) [4–6]. In the case of FCEVs, charging is not required; in BEV, however, it depends completely on external power and requires sufficient power supplied from the grid for further charging. Apart from grid charging the vehicles, they also have accessibility for regenerative braking, i.e., they can charge itself from some of the power they have lost while braking and this adds to its significance.

As clearly mentioned, AEV becomes functional solely by means of electricity. These types of EV have a common efficiency range of 80–100 miles, while in some upgraded version they can last up to 250 miles. When the battery becomes drained, it can take up to one hour (considering fast charging) or even up to a day for normal charging to recharge it again. That also depends on the type of charger and the type of battery used [3, 4].

If the above range is not sufficient, then the PHEV can be used instead. The driver can prefer running on electricity over small ranges (such as, say, 6–40 miles) and then switch over to the internal combustion (IC) type which runs on gasoline when the power has been exhausted. Thus, giving two choices and a result of its flexibility, the PHEV is often to be preferred. Fuel costs are minimized while powering the vehicle from battery and reducing carbon emissions, which is present in the case of a conventional type.

Thus, the PHEV operates on both electric-driven and combustion engine-driven considering long distances, thereby acting like a hybrid EV [6]. It can be powered during acceleration or even when heated, or cooled when air-conditioned. Hydrogen can also act as a fuel cell instead of gasoline in this type of EV.

12.3 OPERATIONAL GUIDELINES AND STANDARDIZATION

With improved EV worldwide, several standards have been made established for dealing with either Institute of Electrical and Electronics Engineers (IEE) or Society of Automotive Engineers (SAE) standards. SAE standards are common in USA manufacturers whereas IEC is most often used in Europe. Similarly, Japan also has its own standardization, Charge de Move (CHAdeMO) and in China there are "Guobiao" (GB/T) standards, which are issued by the Standardization Administration of China, the Chinese National Committee of International Organization for Standardization (ISO) and the International Electro-Technical Commission (IEC). These standards have been adopted globally for charging connectors in connection with the charging of e- vehicles (mainly four-wheel drive vehicles). In this system, the connectors act as an interface between the charging station and the vehicle to be charged. This is similar to the IEC standards. The SAE and the IEC have the same purpose, but use different terms i.e., the SAE describes the level of power as 'purpose' and the IEC refers to it as 'mode' [4, 5].

(1) IEC Standards
   The IEC is an international standardization organization for developing standards commonly for electrical, electronic, and other related technologies (a sector altogether referred to as electro-technology).
   (a) 61851 – It provides a standard for all EVs systems and applicable for both on-board and off-board equipment for charging EVs/PHEVs with supply voltage up to 1000V AC and 1500V DC.
   (b) IEC61980 – It provides a standard for WPT system and is applicable for WPT system with a supply voltage up to 1000V AC and 1500V DC.
   (c) IEC2196 – It provides a standard for conductive charging of EVs like plug, socket, and vehicle inlet.
(2) SAE Standards
   This is a standardization organization mainly based on the USA. It provides standards in different industries mainly focusing on engineering bodies.
   (a) SAEJ2293 – It is applicable for both on-board and off-board equipment for EV charging. It can be classified into two sections:

i. J2293-1 for dealing with how much power is required and maintaining three operating conditions. (i.e., conductive AC, conductive DC, and inductive charging) of the system architecture.

ii. J2293-2 for dealing with required communication and network architecture.

(b) SAEJ1772 – It maintains equipment ratings for charging EV like current rating of circuit breaker, voltage rating of charging etc. It is both applicable in either AC or DC. DC charging maintains higher speed of charging compared to AC (less than 30A). However, the charging rate is restricted by some factors or infrastructure.

(c) SAEJ1773 – It provides a standard for minimum requirement mostly for inductively coupled scheme. It also maintains requirements of software interface and manually connected inductive charging systems [6, 7].

12.4 OPERATIONAL PARAMETERS: MODELLING ASPECTS, MAJOR DRIVES, TECHNICAL PARAMETERS, OPERATIONAL GUIDELINES AND STANDARDIZATION

12.4.1 Modelling aspects

Smart grids can be modelled in real time which improves planning of the system and helps in reducing possible damage. By implementing real-time monitoring in hardware allows putting the test of control algorithms and advanced equipment in smart grids. Wide area monitoring and control (WAMC) test is one such type. In practice, some more types can be mentioned as phase measurement units (PMUs) and the real-time digital simulator (RTDS) (Dorotić, 2019; Naik, 2019). WAMC can protect the smart grid from many interferences considering its reliability, dependability and interrelate control mechanisms to lessen larger area interruption. PMUs are widely used across the globe in view of their precise measurements and controlling aspects which can reduce huge area interference. These PMUs are purely time-based that can make precise decisions and faster response in controlling.

The function of the RTD is for modelling purposes and simulating the power system while PMU give the recorded data for monitoring it. The purpose of WAMC is to monitor the control of algorithm and the system execution.

12.4.2 Major drivers

Today, given the advances in technology, the problem of deteriorating fossil fuels and the decarbonization of the electricity supply are considerably improved with integration of natural resources with new findings, thereby providing a sustainable service to the consumers. A smart grid is an electricity network that allows for a two-way flow of electricity and data, as well as the detection, reaction, and prevention of changes in usage and other issues, using digital communications technology.

12.4.2.1 Components of the smart grid

To attain technologized smart grid, a wide range of automations are complemented, which are briefly discussed below:

(1) Intelligent appliances
    Such appliances decide how much energy are needed, based on the consumer's requirements which are set in advance. This generally helps in aligning electricity generation with the decrease in peak loads. Among the types of appliances in use are smart sensors, as used, in particular, in thermal power stations, which control the temperature of the boiler by observing the previously set values of temperature [7, 8].
(2) Smart power meters
    Figure 12.1 shows the components of smart grids. It is a two-way mode of communication, i.e., between service providers and their respective consumers. Smart meters measure and record the amount of

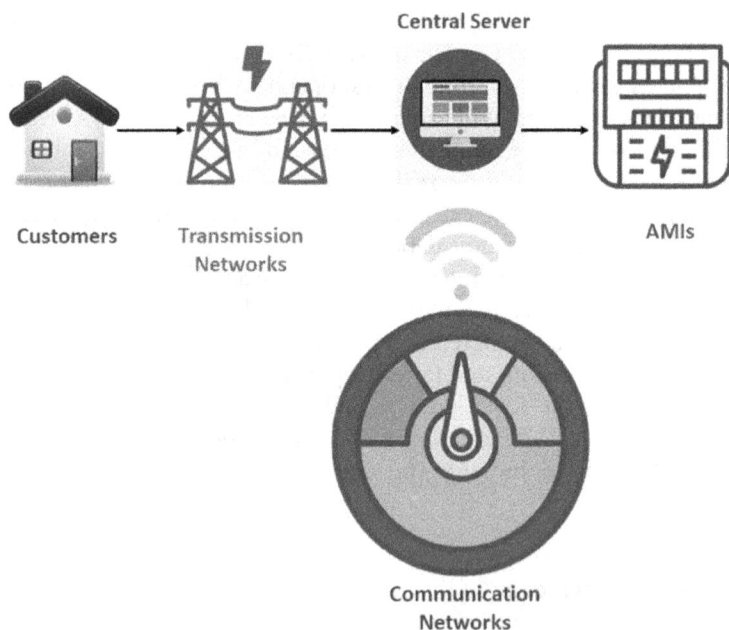

Figure 12.1 Components of the smart grid.

electricity used every 30 minutes. The recorded data are again sent to power distributors who manage the grids and power lines. Data for electricity usage are made available to us via web portal or home display by the providers, which can be used for comparing offers and managing costs. It automatically displays the total billing of data, traces any defaults in the device and forward required services for correction as quickly as possible.

(3) Smart substations

This includes the analyzing and operation of both critical and non-critical parameters such as the circuit breaker, transformers, relays, capacitor banks, and the performance of secure power factors through ensuring the supply of safe and authentic energy. They are also responsible for cleaving the energy directions into several paths. They are operated over a large area and the equipment provided is usually expensive in case of substations as used, for example, in controlling the above-mentioned parameters [9].

(4) Superconducting cables

These are mainly used for transferring long-distance power and the self-monitoring and self- detection of any fault and can forecast the failures of cables in advance, based on real-time weather data and its pre-set outage data.

(5) Integrated communications

Figure 12.2 shows the application of a Supervisory Control and Data Acquisition (SCADA) system in integrated communications. This is the key to a smart grid system. It should be able to respond on a real-time basis and many controllers such as a Programmable Logic Controller

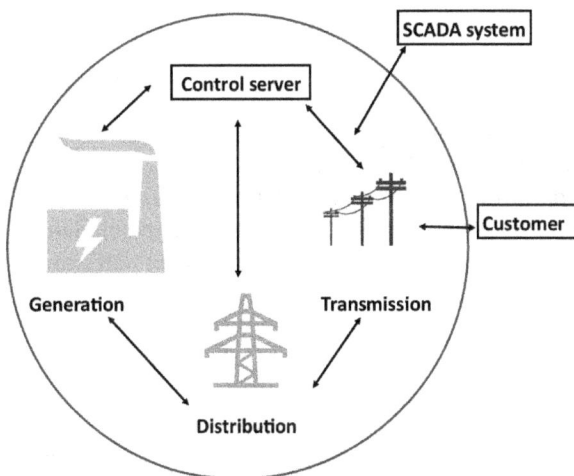

Figure 12.2 The application of the SCADA system in integrated communications.

(PLC) or SCADA are introduced in these communications, depending on our requirements.

In this type of system, we should consider simple stationing, inactivity, standards, dependability and network broadcasting capabilities.

Table 12.1 shows the comparison of various technologies and their risk profiles.

(6) Phase Measurement Units (PMU)

This is applicable when measuring electrical signals based on particular time simultaneity. This synchronization in time gives synchronized results in real time for different remote measurement values on the system [7–10].

Among the benefits of using a smart grid are the following points:

- It provides healthier energy management when compared with combining isolated technologies.
- It ensures protection during any management in electrical equipment under emergency conditions.
- It generates higher demand or supply for the consumers.
- It results in improved power quality.
- It helps in reducing the emissions of carbons.
- As demand for energy increases, the system complexity increases with advanced and improved management.
- Integration with renewable sources improves reliability.

Smart grids have an important role to play in the advance of smart technologies. Some of its applications are given in Table 12.2.

12.4.3 Applications of energy management

12.4.3.1 Power flow

Focusing on the direction of power flow, EV chargers can be categorized as either unidirectional or bidirectional.

For charging in the case of a unidirectional charger, a diode rectifier and a unidirectional DC-DC converter is used for control. This is mainly preferred due its comparative simplicity. It reduces the deterioration of the battery and has lower issues in interconnection when compared to that of bidirectional types [10].

Its major drawback is that it cannot ensure accompanying services to most of the grid.

By contrast, in case of a bidirectional EV charger, a bidirectional DC-DC converter and a bidirectional grid connected AC-DC converter is used, which can give adequate ancillary services as it can act either for discharging or charging operation.

Table 12.1 Comparison table of technology and risk profile

Technology	Deployability	Cost-capital	Cost-ops	Inactive	Speed	Regulatory	Standards	Coverage
Wireless								
Cellular	H	L	H	H	<100 Kbps	L	L	L
900 MHz	M	L	L	M	<1 Mbps	L	L	M
Wifi/WiMAX	L	M	L	M	2–30 + Mbps	L	M	M
Licensed	M	H	M	M	2–30 + Mbps	M	L	M
Microwave	M	H	L	L	10–500 Mbps	L	L	H
Wired								
PLC	L	L	L	M	<100 Kbps	L	M	M
DSL	M	L	M	M	<3 Mbps	L	L	M
BPL	M	M	L	M	2–30 + Mbps	M	H	M
Fixed time	M	L	H	L	2–30 + Mbps	L	L	H
Fibre	H	H	M	L	>Gbps	L	L	H

Table 12.2 Applications of smart grids

Future implementation	Present market
Trading and client service	Flow of data from end to user energy management operations.
Smart charging of vehicle to grid (V2G) and PHEVs	Application data sharing of PHEV
Distributed generation and storage	Analysing of distributed benefits
Grid enhancement	Protects from faults and self-alleviate, manage energy outage, weather data integration, advanced sensing
Response to demand	Load forecasting, advanced demand response
Advanced metering infrastructure (AMI)	Access of remote-meter reading, enhanced security like theft detection, mobile workforce management

Its major drawback is that it causes a reduction in the lifetime of an EV battery due to recurrent cycling of discharging power returning to the grid [9–12]. There's only a limited number of wall boxes on the market which means a lack of competition in price and choice. Above all, it also reduces the durability of the EV through the continuous cycle of power discharged from the grid [9–12].

12.4.3.2 Energy management in the smart grid

As the population increases, the demand of energy also increases and power generation also needs to be increased. As consumers increase, the system load also becomes unpredictable. Accordingly, the unpredictable nature of the load imposes a threat on power utilities and also system operators. Thus, there is uncertainty of high peak load and might lead to a dysfunction in the system parameters [13]. This, in order to cope with this threat, the system operators proposed two methods:

- By increasing the size and range of network, which is very expensive and also its implementation, which is time-consuming.
- The employment of energy management so that the high peak load is minimized during congestion hours.

The importance of energy management has become increasingly significant and is associated with advanced algorithms and ways to manage energy. Thus, it is necessary to introduce a smart grid for the following reasons [12, 13]:

- It is automatic in nature and independent from human interference.
- The results are very accurate and provide forecasting.

- It increases the optimization in generating units and is also cost-effective.
- It protects against power losses on transmission lines and networks, thereby benefitting the system operator with reduced costs of power distribution.
- It provides increased efficiency.
- It helps in the conservation of resources.
- It promotes a greener future by replacing machines that emits greenhouse gases with smart meters with effective reduction in pollution.

a. Applicability

Energy management can be discussed from two distinct points of view: that of the supplier and that of the consumer.
- The energy supplier either the system operator or production units might operate low-cost generators for meeting power demand to consumers through advanced energy management while the higher costs are left for providing extra load demands during times of peak load. Thus, Energy management helps in minimizing the costs of generation units.
- The system operator uses energy management for minimizing losses on lines (either transmission or distribution) and networks and focusing on renewable energy sources (like solar cells/photo voltaic and wind farms, etc. in a more systematic manner) [12].
- The consumers, on the other hand, in the case of industries/households/residential can reduce their electricity bill and planned the load demand with the help of this energy management in an efficient manner [6–8].

b. Tools for energy management

The generating units and system appliances are often controlled through the use of basic controls or through manual operation. With the improved technology, the computer-based system and the introduction of the advanced algorithm, it becomes more convenient for the system operators to control the load. Among the most commonly used control techniques are:
- PLC (Programmable Logic Controller)
- SCADA (Supervisory Control and Data Acquisition)
- EMS (Energy Management System)
- Automation systems (like some home automations)

c. Energy Management in the Tariff System

The above points show some of the control techniques used both in hardware and software control. Software control is usually done with the help of engineers and software specialists. The hardware is generally composed of inputs and outputs where the system can function as either on or off because of some connected device or change in demand for power.

It can be concluded that the objectives of energy management are to reduce the economic losses and costs as much as possible. In order to fulfil this requirement, however, there need to be some changes in the electricity tariff system. In most countries, a common tariff is usually used, such as a tariff of kWh kept constant for different timings in a day. Another system is that the tariff increases respectively with the energy consumption which has been the practice in some countries [10]. Thus, several tariffs have been introduced such as Demand Response Programs (DRPs), where electricity tariffs are variable depending on the time. In addition, those consumers who are found to consume more than limits against the tariffs can be penalized.

d. Issues faced in the current scenario

The improvement and implementation of new technologies are one of the major challenges. Energy management is one of the most significant of these current developments. There are, however, certain criteria which limits this implementation.

Some of the limitations are listed below:

- It is not cost-effective, and thereby faces a major challenge in its implementation. Moreover, the profit on investment is very low compared to the installed cost.
- It becomes difficult both for electric utilities and retailers when the system of electricity tariffs is changed frequently. It should be made changeable along with the time.
- The system operator used to be faced with huge investments while trying to upgrade the network infrastructure.
- The bidirectional flow of power is as yet to be put into practice, which has caused a delay in terms of ideal energy management.
- Comprehensive knowledge and background of the population is essential for securing rapid operation.
- Energy management has been introduced through the process of upgrading from traditional to newer and smarter methods, focusing mainly on global warming [14, 15]. But in some cases, it also has some negative impacts on humans. For example, the replacement of previously existing analog meters with high-cost electronic meters.

12.5 MAXIMIZING RESS UTILIZATION

Renewable energy can be defined as energy which is available in nature and which can be repeatedly generated from natural phenomena as it is capable of being replenished. Among the most examples of renewable energy are sunlight, air, water, heat, biomass, geothermal, etc. It is commonly applicable in four important areas: air conditioning (cooling/heating), electricity generation, small projects in rural areas and in various fields of transportation [7–10].

Currently, people have become increasingly aware of climate issues, due to increased reporting global warming and the greenhouse effect. Accordingly, they have started to invest their money and to utilize these resources, recognising its significance in promoting sustainable development.

From a recent study of REN21 (Renewable Energy Policy Network for the 21st Century), which is a global policy network of committed individuals working to create a sustainable renewable energy future, the consumption of renewable energy has been increasing rapidly over the past few years. It has been used for global electricity production, which is found to be relatively small all across the globe, i.e., it was only 27.3% in 2019 and increased to little more than 30% in 2021.

The importance of renewable energy has increased all over the world considering confined production of conventional type energy resources in specific geographical areas only. One example is that much of the world's gas and oil is produced in the Middle East. The practice of RESs utilization also helps in reducing pollution to a great extent and also plays an important role in the development of countries' economies and increasing energy security [5].

12.5.1 Importance of the maximization of RES

First, it is eco-friendly and lessens the pollution in an effective way while it powers millions of homes and industries. A large number of wind farms have now been erected and solar energy can be utilized through their installation on roof tops widely all over India. But the number of installations is relatively low by global standards. We are still in a race to search for alternatives to clean energy. The National Research Development Corporation (NRDC) are trying hard to extend these installations and to benefit from these resources.

Several policies have been framed in the USA, with the intention of improving the net production of wind and solar energy, and promote net metering. In other words, ordinary people installing solar panels can sell the excess power back to the grids or any other companies and benefit from it. They support incentives that enhance renewable energy and aim to increase the energy production from wind and solar to a target of 30% of US total energy production. They try to connect a nationwide transmission of grid and contribute to making the world a cleaner environment. They also attempt to mitigate the impacts on wildlife when installing wind and solar plants [7–10]. India and China have also introduced initiatives to install wind and solar plants in an efficient manner in order to reduce environmental pollution.

In China, the NRDC has taken up the initiative for introducing a renewable power grid that can withstand a high penetration of electrical energy.

In India also, the government should take up actions in order to introduce wind and solar energy and provide financial structures for making the earth a cleaner and healthier planet [12–14].

In Latin America, the NRDC, along with some groups of partners, have begun to propose that governments in the region should increase research into renewable sources instead of continuously depleting their fossil resources.

12.6 ENHANCING OPERATIONAL EFFICACY MEASURES AND METHODS

There are many opportunities provided when looked into a vast scope in smart grid technologies with regard to considering its operational efficacies.
The benefits provided can be discussed as [9]:

(1) Coherent use of capital investments.
(2) Systematic real-time operations.
(3) Systematic maintenance and management.

Some of the operational efficacy of smart grids can be discussed briefly.
Most of the smart grid's aims are focusing on providing efficient service in electricity. The benefits we get from this efficiency can be discussed broadly in these specific areas:

- Investment expenditure in the areas of transmission, generation and distribution.
- Instantaneous operations in the generation of electricity, additional service and the depletion of equipment.
- Maintenance and management costs, the meeting of customer requirements, failure recovery and metering.

The costs classification and profits gained are given in Table 12.3.

12.7 ECONOMIC ASPECTS, AFFORDABILITY AND PRACTICABILITY

The power system implicating smart grids have benefited directly both economically and environmentally, influencing both the service provider and the operator. Smart grids provide ancillary services and the integration of renewable sources. The initial investment is costly; once installed, however, the benefits are more than repaid with a gradual reduction in operating costs. This is used for two-way communication purposes where the related customer has to face a great expense in times of initial investment, but the

Table 12.3 Costs classification and profit gained

Capital investments	Real-time operations	Maintenance and administration
1. Generation capacity for basic and peak demand, and ancillary services • Regulation • Spinning • Reserves • Non-spinning reserves 2. Transmission capacity for peak demand • right-of-way & construction • Reconstructing / voltage upgrades 3. Distribution capacity for peak demand • substations & transformers • feeder & branch circuits • customer transformers • voltage regulator • reconstructing/ voltage upgrades	1. Fuel to generate electricity. 2. Wholesale purchases/sales of electricity. 3. Ancillary service operations costs • Regulation • Spinning reserves • Non-spinning reserves 4. Equipment wears and tear • Generators • Transformer • Circuit breakers, etc. 5. Electrical losses in T&D 6. Fault location and isolation	1. Distribution control operations • feeder reconfiguration • load balancing 2. Maintenance labour • Generation • T&D 3. Distribution outage management/ restoration 4. Meter reading 5. Service connection / disconnection 6. Theft of electricity 7. Customer service 8. Billing

costs diminish in the long run [9, 10]. The need for generation, transmission and distribution in the electricity market will become reduced in future with the gradual technological advances, which has an impact in terms of reducing capital costs in the near future. But this ensures the operation and maintenance costs will also be reduced. As mentioned above, the advantage of operating a smart grid might be due to savings from reduced peak load, reduced in-line losses or reduction due to optimization and artificial intelligence and reduction in economic loss (in case some parts of the population are exempt from metering). The profit will also increase in line with increases in system reliability, the best performance of power factor and improvements in system security. With the use of increased renewable energy sources, the environmental status also improves as emissions of carbon dioxide causing global warming are reduced, as are also several other pollutants.

More areas profiting the system operators can also arise from improved technologies, like leap-frogging, minimizing the risks in the system and supply shortfalls and producing products of smart grid and monetizing it in the global market. It also enables the establishment of new industries, providing jobs in that field, enthusiastic young minds and conserving a huge platform in the fields of trading.

12.8 OPTIMAL ECONOMICAL MODEL

The modern advancements of modelling and simulations of science and technology have been solving complex issues for clients across various industries. Figure 12.3 demonstrates a smart grid where independent renewable energy sources are integrated with conventional energy sources in order to provide energy to a mine. The model exports financial and operational statistics to the user, such as the labelled cost of energy, energy shortfalls as well as to statistics pertaining to day to day or annual energy source performance. This model can be used to find the optimal configuration of available energy sources and storage batteries to reduce the costs of overall providing energy for the mine [7–11]. The amount of energy solar panels are capable of providing each cycle depends on the insulation during that time of day or year. Any energy source can be sized or turned on and off in the user interface. For the operation, energies are dispatched according to our desired time in the mine.

There are also a number of shortfalls in such a system. Operational impacts have reserved battery sizes as well as source-specific statistics such as generator running, standby and off percentages and daily fuel consumption. It focused on insisting that clients make better decisions through the applications of decision support tools. It is associated with the team having

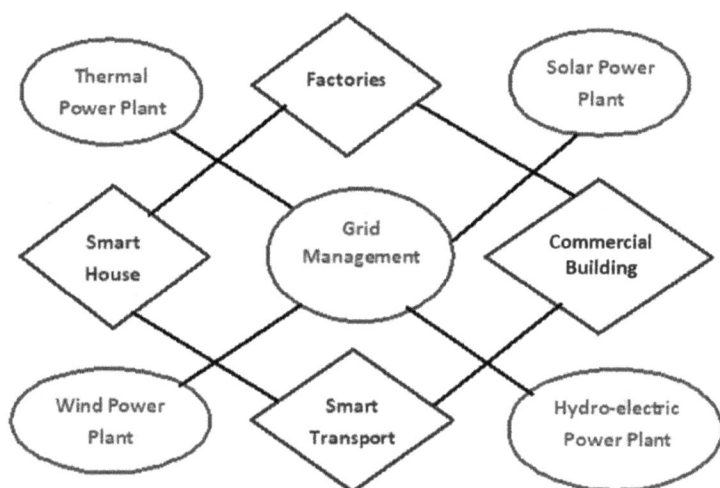

Figure 12.3 Economical model of a smart grid system integrated with conventional energy sources.

extensive experience in working with leading computer simulation software and solving complex issues for clients across various fields.

12.9 COST-AWARE EVS INTERACTING WITH RESS

The integration of EVs with renewable energy sources can significantly reduce the barriers to smart grids and allow the generation of more electricity using these resources. Compared to that of conventional plants, RERs are more reliable and predictable and also more complex to control due to small impact of size, since they are integrated compactly, with large numbers requiring an improved market platform. The implementation of a smart grid system is associated with demand response, genetic algorithm, controlling software and balance on both the supply and the demand side. With this technology, the operators can easily control the grid in any response conditions [12–14]. This helps in influencing the integration of grids with that of RESs in a more efficient and cost-effective manner.

When distributed generators are installed in association with some distributed system management, then the initial capital costs are considerably reduced in power grids with efficient management in its controls and operations.

Smart grid technologies allow multiple change in renewables such as transmission and distribution systems, self-operating and active distributed energy resources, to be assisted on the grid. When an infrequent resource is no longer capable of producing energy for a long period of time, a renewal of energy is required. This replacing is usually supplied by the power plants operating on large-scale, which can scale up and down until the desired power balance is attained. These balanced power plants can also produce energy from conventional type resources, such as hydro plants, natural gas and biofuels as well as biomass [13–15].

The replacement of fossil fuels by these renewable energy sources has a substantial impact on carbon emissions. With the reduction in required reserve capacity, the Pacific Northwest National Laboratory (PNNL 2010) reported that the 25% of wind penetration in the US has been reduced by 5% in terms of consumption of electricity. That reduced percentage is associated with a 5% reduction in CO_2 emissions. The indirect reductions are based upon assumptions that the capital costs funded for the installation of fossil fuel plants are invested on renewable generating units instead.

12.10 EMISSION-AWARE EVS INTERACTING WITH RESS

EVs help in keeping the environment free from pollution and mostly reduces the carbon emissions in the atmosphere. This, in turn, has a great impact on climate change and smog compared with conventional vehicles.

Generally, emissions from vehicles are classified into two types: direct and life cycle.

1. Direct emissions: Direct emissions come from the tailpipe of vehicles and are produced directly during the fueling process. The emissions include smog components which are hazardous to health and are mainly considered as GHGs, principally a high CO_2 content. Electric vehicles are mainly used given their zero-emitting nature in cities and urban areas. PHEVs are associated with gasoline in addition to the electric motor, which also produce emissions from fuelling. They are, however, more environmentally friendly when compared to the conventional vehicles which produced more carbon emissions [17].
2. Life cycle emissions: In this type, all emissions of gases relating to vehicle production, exhaust, fueling, recycling etc. are included. For example, the common gasoline which we use in vehicles is extracted from the earth, refined, transported to gas stations and then ignites inside the vehicles. Similar to direct emissions, life cycle emissions also emit several harmful pollutants and GHGs [15–17].

Almost all vehicles produce emissions and calculating the quantity they produce is no easy task. But the emissions from EVs are considerably lower than the conventional ones as they are associated with electricity generation rather than fuel ignition. The amounts of EVs' emissions are varying with different geographical areas. EV drivers can further improve the emissions and reduce the life cycle by integrating with environmentally-friendly resources like that of wind and solar energy.

12.11 CASE STUDIES

Table 12.4 shows the case studies of different electric vehicles.

12.12 MATHEMATICAL MODEL

As the demand of electricity continues to increase, the system reliability and cost functions needs to be improved in an efficient manner. Based on different aspects and controlling methods, several objective functions have been adapted so as to obtain global optimization solutions for different electrical users and various applications in EVs.

Among the techniques adapted are coalitional game theory, tree- based methods, fuzzy logic control, dynamic programming, genetic algorithm, equivalent energy consumption minimization strategy, and modern predictive control. These techniques help in optimizing the control parameters in order to achieve the desired output.

Table 12.4 Electric vehicles case studies

Case study	Characteristics	Main results	References
Electric buses, Guwahati city, Assam, India	A very high quantity of energy storage medium is implemented between the grid and the transportation system. Introduction of a solar plant for reducing system dependence requirement on the grid.	Number of internal combustion engines are reduced and results in better air quality. Integrating with renewable energy sources provides support to the electrification. State of transportation is completely transformed.	[16]
EVs applications at the Caribbean Island of Barbados	Peak electricity demand is 167.5 MW and generation range is 240 MW. A 10 MW PV for utility and 20 MW of distributed solar PV on buildings and commercial rooftops. They can be considered either as static or a smart charging device with V2G capability.	Storage capacity gets reduced to 13% when EVs are charged during daytime. V2G helps in increasing renewable penetration and reduces initial investments in grid-connected storage up to 12% when charged during night time and 20% during daytime.	[17]
São Miguel Island, Azores, Portugal	Analysis of medium voltage distribution network. Generation of an oil immersed transformer supplying one private industry at 250 kVA, 10 kV/0.4 kV along with an analysis of EV charging scheduler.	With 60% EV penetration transformer efficiency improves at a range of 7.54% and up to 76% at 100% EV penetration. Addition of EV charging scheduler improves 98% of life.	[18]

(Continued)

Table 12.4 Electric vehicles case studies (Continued)

Case study	Characteristics	Main results	References
Tenerife, Spain	EVs penetration is analysed from the data obtained from cars, mobility and power demand.	EVs can store and reduce difference in amplitudes between values and peaks of the electric energy demands curve from penetration and can handle and manage smart grids in an efficient manner.	[19]
Korcula, Croatia	Analysis of 100% renewable energy sources (solar and wind) in combination with 100% share of vehicles is performed	The import and export peak loads have no effect from the smart charge share reduction of EVs.	[2]
UCLA campus city, Santa Monica	UCLA Smart Grid Energy Research Centre (SMERC) is used as a test bed. Performing optimization at two levels: (a) to find the optimal EV load profile for the day-ahead energy market; (b) to control all online EVs in the system so that optimal load profile and charging scheme is applied from stage (a). EV users' patterns are analyzed for optimizing the EVs operation and coupled to a demand response program.	With lowering of computational costs of the system, 18% energy cost is reduced.	[20]

Table 12.5 shows the different mathematical approaches/techniques used for the optimal planning of EVs.

12.13 FUTURE DIRECTIONS

With the considerable increase in electricity demand in different regions of the nation, there is a desperate need to implement the system of a smart grid. The introduction of a smart grid requires a great deal of cooperation

Table 12.5 Mathematical model

Sl no.	Objective functions	Optimization technique/methods	Remarks	References
1	Coalitional game theory		Used in selection of optimal selection of electricity consumers for participating in DR schemes, and the allocation of the coalition's payoff among DR participants (known as solution concept).	[21]
2	Tree-based methods	Power consumption of a building as a function of the temperature, humidity, wind, time of day, type of day, schedule, lighting level, water temperature and historic power consumption.	Uses regression trees for modelling of energy consumption regarding cooling systems and compares the outputs with SVR methods.	[21]
3	Fuzzy logic control		Used in minimizing/maximizing of cost function over a fixed drive cycle of a fuel cell EV	[16]
4	Dynamic programming		Used in optimizing power for known drive cycle and simplifying a complicated problem by breaking it into simpler sub-problems in a recursive manner.	[17]

(Continued)

Table 12.5 Mathematical model (Continued)

Sl no.	Objective functions	Optimization technique/methods	Remarks	References
5	Genetic algorithm	It starts with a set of solutions (chromosomes) called a population. The solutions from one population are taken according to their fitness to form new ones. Most suitable solutions will get a better chance than the poorer solutions to grow and the process is repeated until the desired condition is satisfied.	Can generate solution to optimization and search problems.	[18]
6	Equivalent Consumption Minimization Strategy (ECMS)		Reduces the global optimization problem and calculates total fuel consumption as sum of real fuel consumption by ICE and equivalent fuel consumption of electric motor at each instant.	[19]
7	Modern Predictive Control (MPC)	$\{u(k + i - 1), i = 1, 2, ..., M\}$ $y^\wedge (k + i), i = 1, 2, ..., P\}$	It can predict the output behaviour then, optimize the control parameters to achieve the desired output.	[20]

with clients in this power-generating system. The shortage of power due to an increased number of consumers is a common scenario usually found in India. The generated electricity grids could not provide sufficient capacity or system security, are less efficient, less reliable and have a great impact on the existing environment as well in trying to meet the electricity requirements of modern society. Therefore, it is high time for us to bring our country to a revolutionized society. And that is possible through the introduction of a smart grid. This will surely lead us to a new different path in the process of development [16–18].

The Electricity Advisory Committee (EAC) states that considering the importance of smart grid, it is of particular importance for the nation to practice the cost-effective stationing of smart grid techniques [19]. It can be a means to achieve the nation's goal in areas of security of supply, changes in the climate, grid dependability, the growth in economy and nation competitions globally. But there are still a number of challenges associated with the development of a smart grid. The EAC proposed the following suggestions to the US Department of Energy (DOE):

1. To create a smart grid programme office within the DOE.
   The office should maintain:
   a. Clearing of worldwide smart grid information through internet servicing tools.
   b. Information regarding smart grid business schemes, and effective regulatory models.
   c. Education on stakeholders, utility regulators and other consumers.
   d. Coordination between several smart grid organizations.
   e. Proper standards as soon as the National Institute of Standards and Technology (NIST) completes the development of a planned frame, which is approved in Section 1305 in the Energy Independence and Security Act of 2007 [20].
2. To develop, control and communicate R&D development projects which is compatible with the roadmap, and also directs the Smart Grid Regional Demonstration Initiative and Matching Grant Program as authorized in EISA 2007 and referenced above.
3. To conduct an important educational campaign session based on enlightening the consumers regarding the costs of energy and how it can be fully utilized.
4. To establish smart grid engineering and technical development so that the interested students can pursue smart grid-based technical courses. These should be required training at all levels. And also, to create a workgroup for ensuring the interested candidates fulfil the require needs for the programme.
5. To work with industries and state regulators enhancing motivation and creating standards in order to ensure the efficient operation of smart grids.

12.14 CONCLUSION

With the integration of renewable energy sources in EVs, this chapter has laid out several transportation sectors and various power-generating units have dramatically reduced emissions of harmful gases such as carbon dioxide. Large applications of EVs will have a considerable impact and will provide us tremendous benefits with the adoption of these resources to present electrical grids. Operational parameters, including modelling aspects and major drivers, and technical parameters, such as power flow and energy management issues, have been discussed in this chapter and the way in which RESs' utilization can be maximized has also been discussed, along with a case study detailing several present-day practical implementations.

REFERENCES

[1] Lakshmi, G. Sree, et al. "Electric vehicles integration with renewable energy sources and smart grids." *Advances in Smart Grid Technology*. Springer, Singapore, (2020): 397–411.

[2] Dorotić, Hrvoje, Borna Doračić, Viktorija Dobravec, Tomislav Pukšec, Goran Krajačić, and Neven Duić. "Integration of transport and energy sectors in island communities with 100% intermittent renewable energy sources." *Renewable and Sustainable Energy Reviews* 99 (2019): 109–124.

[3] Paget, Mia, Tom Seacrest, Steve Widergren, Patrick Balducci, Alice Orrell, and Cary Bloyd. "Using smart grids to enhance use of energy-efficiency and renewable-energy technologies." *Pacific Northwest National Lab.(PNNL)* No. PNNL-20389 (2011).

[4] Tie, Siang Fui, and Chee Wei Tan. "A review of energy sources and energy management system in electric vehicles." *Renewable and Sustainable Energy Reviews* 20 (2013): 82–102.

[5] Hannan, M. A., F. A. Azidin, and Azah Mohamed. "Hybrid electric vehicles and their challenges: A review." *Renewable and Sustainable Energy Reviews* 29 (2014): 135–150.

[6] Hannan, M. A., Md M. Hoque, Azah Mohamed, and Afida Ayob. "Review of energy storage systems for electric vehicle applications: Issues and challenges." *Renewable and Sustainable Energy Reviews* 69 (2017): 771–789.

[7] Singh, Mukesh, Praveen Kumar, and Indrani Kar. "A multi charging station for electric vehicles and its utilization for load management and the grid support." *IEEE Transactions on Smart Grid* 4, no. 2 (2013): 1026–1037.

[8] Arancibia, Arnaldo, and Kai Strunz. "Modeling of an electric vehicle charging station for fast DC charging." In *2012 IEEE International Electric Vehicle Conference* (2012): 1–6.

[9] Yilmaz, Murat, and Philip T. Krein. "Review of battery charger topologies, charging power levels, and infrastructure for plug-in electric and hybrid vehicles." *IEEE Transactions on Power Electronics* 28, no. 5 (2012): 2151–2169.

[10] Ahmad, Aqueel, Zeeshan Ahmad Khan, Mohammad Saad Alam, and Siddique Khateeb. "A review of the electric vehicle charging techniques, standards,

progression and evolution of EV technologies in Germany." *Smart Science* 6, no. 1 (2018): 36–53.

[11] Hannan, Qi, Hailong Li, Lijing Zhu, Pietro Elia Campana, Huihui Lu, Fredrik Wallin, and Qie Sun. "Factors influencing the economics of public charging infrastructures for EV–A review." *Renewable and Sustainable Energy Reviews* 94 (2018): 500–509.

[12] Tan, Kang Miao, Vigna K. Ramachandaramurthy, and Jia Ying Yong. "Integration of electric vehicles in smart grid: A review on vehicle to grid technologies and optimization techniques." *Renewable and Sustainable Energy Reviews* 53 (2016): 720–732.

[13] Sharma, Angshuman, and Santanu Sharma. "Review of power electronics in vehicle-to-grid systems." *Journal of Energy Storage* 21 (2019): 337–361.

[14] Heydari, Ali, and Alireza Askarzadeh. "Techno-economic analysis of a PV/biomass/fuel cell energy system considering different fuel cell system initial capital costs." *Solar Energy* 133 (2016): 409–420.

[15] Newell, Richard, Daniel Raimi, and Gloria Aldana. "Global energy outlook 2019: the next generation of energy." *Resources for the Future* 1 (2019): 8–19.

[16] Naik, M. Bhaskar, Praveen Kumar, and Somanath Majhi. "Smart public transportation network expansion and its interaction with the grid." *International Journal of Electrical Power & Energy Systems* 105 (2019): 365–380.

[17] Taibi, Emanuele, Carlos Fernández del Valle, and Mark Howells. "Strategies for solar and wind integration by leveraging flexibility from electric vehicles: The Barbados case study." *Energy* 164 (2018): 65–78.

[18] Godina, Radu, Eduardo MG Rodrigues, João CO Matias, and João PS Catalão. "Smart electric vehicle charging scheduler for overloading prevention of an industry client power distribution transformer." *Applied Energy* 178 (2016): 29–42.

[19] Colmenar-Santos, Antonio, Ana-Rosa Linares-Mena, David Borge-Diez, and Carlos-Domingo Quinto-Alemany. "Impact assessment of electric vehicles on islands grids: A case study for Tenerife (Spain)." *Energy* 120 (2017): 385–396.

[20] Xiong, Yingqi, Bin Wang, Chi-cheng Chu, and Rajit Gadh. "Vehicle grid integration for demand response with mixture user model and decentralized optimization." *Applied Energy* 231 (2018): 481–493.

[21] Antonopoulos, Ioannis, Valentin Robu, Benoit Couraud, Desen Kirli, Sonam Norbu, Aristides Kiprakis, David Flynn, Sergio Elizondo-Gonzalez, and Steve Wattam. "Artificial intelligence and machine learning approaches to energy demand-side response: A systematic review." *Renewable and Sustainable Energy Reviews* 130 (2020): 109899.

Index